The Crust
of Our Earth

PHalarope Books

PHalarope Books are designed specifically for the amateur naturalist. These volumes represent excellence in natural history publishing. Each book in the PHalarope series is based on a nature course or program at the college or adult education level or is sponsored by a museum or nature center. Each PHalarope Book reflects the author's teaching ability as well as writing ability.

BOOKS IN THE SERIES

Born and raised in Tennessee, **Chet Raymo** graduated from the University of Notre Dame with a degree in engineering and a Ph.D. in physics. Since then he has taught physics, astronomy, and other science-related topics at Stonehill College in North Easton, Massachusetts. He is the author of the highly acclaimed *365 Starry Nights* (Prentice-Hall, 1982).

THE CRUST OF OUR EARTH

An Armchair Traveler's Guide to the New Geology

CHET
RAYMO

Prentice-Hall, Inc., Englewood Cliffs, New Jersey 07632 A SPECTRUM BOOK

Library of Congress Cataloging in Publication Data

Raymo, Chet.
　　The crust of our earth.

　　(A PHalarope book)
　　"A Spectrum Book."
　　Includes index.
　　1. Earth—Crust.　　2. Plate tectonics.　　I. Title.
QE511.R328　　　1983　　　551　　　83-9511
ISBN 0-13-195107-6
ISBN 0-13-195099-1 (pbk.)

This book is available at a special discount when ordered in
bulk quantities. Contact Prentice-Hall, Inc., General
Publishing Division, Special Sales, Englewood Cliffs, N.J. 07632

10　9　8　7　6　5　4　3　2　1

ISBN 0-13-195107-6
ISBN 0-13-195099-1 {PBK.}

Editorial/production supervision
by Kimberly Mazur
Manufacturing buyer: Edward J. Ellis

Prentice-Hall International, Inc., *London*
Prentice-Hall of Australia Pty. Limited, *Sydney*
Prentice-Hall Canada Inc., *Toronto*
Prentice-Hall of India Private Limited, *New Delhi*
Prentice-Hall of Japan, Inc., *Tokyo*
Prentice-Hall of Southeast Asia Pte. Ltd., *Singapore*
Whitehall Books Limited, Wellington, *New Zealand*
Editora Prentice-Hall do Brasil Ltda., *Rio de Janeiro*

When my three oldest children were young, we visited together many of the places described in this book. Like Pictish kings and queens of old we sat in the Wishing Seat at the Giant's Causeway in County Antrim, Ireland. Together we strained our eyes searching for the dark monster in Loch Ness. We stood on the high white cliffs at Dover and tried to imagine the numberless generations of marine organisms that had contributed their tiny skeletons to those thick beds of chalk. We lay on our backs in the cave at Altamira, Spain, and gaped at ice age deer and bison cavorting on the ceiling of the cavern. Those were happy times, times of high adventure, never to return. To Maureen, Dan and Margaret I dedicate this book and wish them well.

Contents

Preface

I first heard of continental drift in early 1963 when I read an article on the subject in *Scientific American* by the Canadian geologist J. Tuzo Wilson. I was a graduate student in physics at the time, and I was sufficiently versed in science to recognize in Wilson's outline of the new theory the kind of economy and elegance that characterizes the best science. I began to eagerly follow the rapid-fire developments unleashed by tentative acceptance of the new ideas. An interpretation of oceanic magnetic anomalies by Fred Vine and Drummond Matthews later that same year (an inconspicuous little paper in *Nature*) captured my attention. The anomalies, wrote Vine and Matthews, are "consistent with, in fact virtually a corollary of, current ideas on ocean floor spreading and periodic reversals of the earth's magnetic field." This little paper turned out to be the key, the nexus, the turning point. The next few years saw a flood of work on magnetic polarity reversals, transform faults, earthquakes, and deep-sea sediments. Pieces of the geological puzzle fell into place with astonishing alacrity. By 1968 the revolution in earth science was virtually complete and old ideas of continental drift had found a secure place in a new and comprehensive theory of plate tectonics. The voyages of the deep-sea drilling vessel *Glomar Challenger* in 1968–70 seemed to confirm on every side the validity of the new theory. It was an exciting time for geologists—and for those of us who sat on the sidelines and watched, mouth agape, all agog. During no other decade of human history have we learned so much so fast about the planet we live on.

Nor has the pace of discovery slowed. A thrill of excitement still runs though the geological community. This sort of speculative fever can be the source of a few problems for a book such as this one. First, plate tectonics has become the new orthodoxy, applied universally and unabashedly to the crust of the earth even where the facts are few and the data skimpy. There are dissenters within the geological community, those who urge caution, and those who are quick to point out inconsistencies and anomalies in the application of the new theory. And rightly so. Let the reader be forewarned that the explication of the earth's crust in terms of the new geology is not always so tidy as it may seem in the pages that follow. Second, the continuing pell-mell pace of discovery means that it is almost impossible for a book such as this to be absolutely up to date. Even as these pages go to press, some of the topics described herein are subject to reinterpretation.

There have been a number of books explaining the new geology to the lay person. This book is different in several respects. The presentation is equally visual and verbal. In a sense, the text can be considered as a kind of extended caption for the illustrations. Reflected on almost every page is my own love for maps. This book is not so much an explanation of the new geology as it is an armchair traveler's appreciation of the crust of the earth in terms of the new geology. This is a journey of the imagination from inky-dark hot-water vents on the floor of the Pacific right around the earth to the glistening ice cap of Antarctica. Along the way we shall see how familiar features of the earth's crust were shaped by the slip and shove of moving plates.

The maps and diagrams of the book presuppose that the reader has some basic knowledge of geography. I have avoided excessive labeling on the maps for fear of cluttering the scientific content. If your geography is weak, you may want to read the book with an atlas at hand. I have also tried to avoid technical language. The Introduction develops what minimum technical background we shall need for our journey. Where cross-referencing might help the reader to understand difficult concepts, I have placed the referenced essay numbers in brackets. There is a glossary of technical terms at the back of the book.

On page xv there is a list of some of the sources I relied on closely for my illustrations. A hearty, though nameless, thank you is here proffered to authors of the many other books and articles that provided inspiration for this around-the-world geological tour. I hope geologists everywhere will take satisfaction in my attempt to bring something of the beauty and excitement of their work before a wider audience.

I would like to thank my daughter, Maureen Raymo, herself a geologist, for reading the text and providing many helpful suggestions, and my wife Maureen for taking the armchair journey with red pencil in hand. Robert Kruse, Academic Dean at Stonehill College was generous in his support of this project. Susan McGrath provided an assist at a crucial juncture and is remembered here with love. Thanks also go to Mary Kennan, editor in the General Publishing Division at Prentice-Hall, and Kim Mazur and Maria Carella, also in GPD, for having confidence in this project and for skillfully giving it a handsome form. And finally, a promised thank you to Elizabeth, John, and Sean Devane for helping to provide a wonderful place in which to write and draw *The Crust of Our Earth*.

Chet Raymo

Notes and Acknowledgments

Hundreds of books, texts, maps, and journal articles were consulted while planning this armchair journey. Some particularly close debts are acknowledged here by essay number:

2. The earthquake eyewitness report is from Gordon Thomas and Max Morgan Witts, *The San Francisco Earthquake*. New York: Stein and Day, 1971.

4. As aids in making the bird's-eye view of Southern California I used the splendid plastic topographic relief maps available from Hubbard, Box 104, Northbrook, Ill., 60062, and the Geological Highway maps produced by the American Association of Petroleum Geologists, Box 979, Tulsa, OK. 74101.

5. My discussion of the origin of the Grand Canyon closely follows John S. Shelton, *Geology Illustrated*. San Francisco: W. H. Freeman and Co., 1966. This book is illustrated with hundreds of aerial photographs taken by the author and beautiful interpretive drawings by the author's brother. It's the next best thing to being there. The sketch of the Grand Canyon is adapted from a U.S. Geological Survey photograph.

6. My sketch of Meteor Crater is adapted from a photograph by Shelton (see **5** above).

7. My sketch of the Lake Bonnevile Terraces is adapted from a drawing in G. K. Gilbert, "Monograph Number 1", U.S. Geological Survey.

10. As a guide in making the geological cross-section of the Rocky Mountain region I used the Geological Highway Maps of the American Association of Petroleum Geologists (see **4** above).

13. The map showing historic earthquakes is adapted from *Earthquake History of the United States*, U.S. Department of Commerce, 1973.

16. The maps illustrating this essay are based on the classic work by J. L. Hough, *Geology of the Great Lakes*. Urbana, Il: University of Illinois Press, 1958.

17. The drawings illustrating this essay are closely adapted from one of the most delightful little books ever written on geology: Arthur N. Strahler, *A Geologist's View of Cape Cod*. New York: Natural History Press, 1966. Adapted with permission of Doubleday & Co., Inc.

18. The illustrations for this essay are adapted from A. Hallam, "Continental Drift and the Fossil Record," *Scientific American*, 227, no. 5 (Nov. 1972), pp. 56–66.

21. The drawing of Greenland rocks is adapted from a photograph by Stephen Moorbath illustrating Stephen Moorbath, "The Oldest Rocks and the Growth of Continents," *Scientific American*, 236, no. 3 (Mar. 1977), pp. 92–104.

26. The map of the North Atlantic basin is based on the magnificent sea floor maps prepared by Bruce Heezen and Marie Tharp. Marie Tharp, oceanographer, has been associated with the Lamont-Doherty Geological Observatory since its founding in the late 1940s. During this time she collaborated with the late Dr. Bruce C. Heezen in the study of sea floor topography. The world ocean floor panorama is based upon soundings of continually increasing accuracy from many sources during a time of changing concepts of the geology of our earth. This chart is available in several sizes from Marie Tharp, One Washington Avenue, South Nyack, New York 10960.

43. The drawings illustrating this essay owe a close debt to illustrations in Haroun Tazieff, "The Afar Triangle," *Scientific American*, 222, no. 2 (Feb. 1970), pp. 32–40.

44. The drawing of the cleft-back antelope is adapted from an illustration in the imaginative book by Dougal Dixon, *After Man*. London: Harrow House, 1981.

45. The illustrations for this essay follow Richard L. Hay, and Mary D. Leakey, "The Fossil Footprints of Laetoli," *Scientific American*, 246, no. 2 (Feb. 1982), pp. 50–57, and Mary D. Leakey, "Footprints in the Ashes of Time," *National Geographic*, 155, no. 4 (Apr. 1979), pp. 446–57.

The map showing oil and gas deposits in the Persian Gulf is based on information in *Rand McNally Atlas of the Oceans*. New York: Rand McNally and Co., 1977.

52. The map of the Pacific basin near the Tonga Trench is adapted from the sea floor maps of Bruce Heezen and Marie Tharp (see **26** above).

54. This essay and its illustrations are based closely on David I. Groves, John S. R. Dunlop, and Roger Buick, "An Early Habitat of Life," *Scientific American*, 245, no. 4 (Oct. 1981), pp. 64–73.

The Crust
of Our Earth

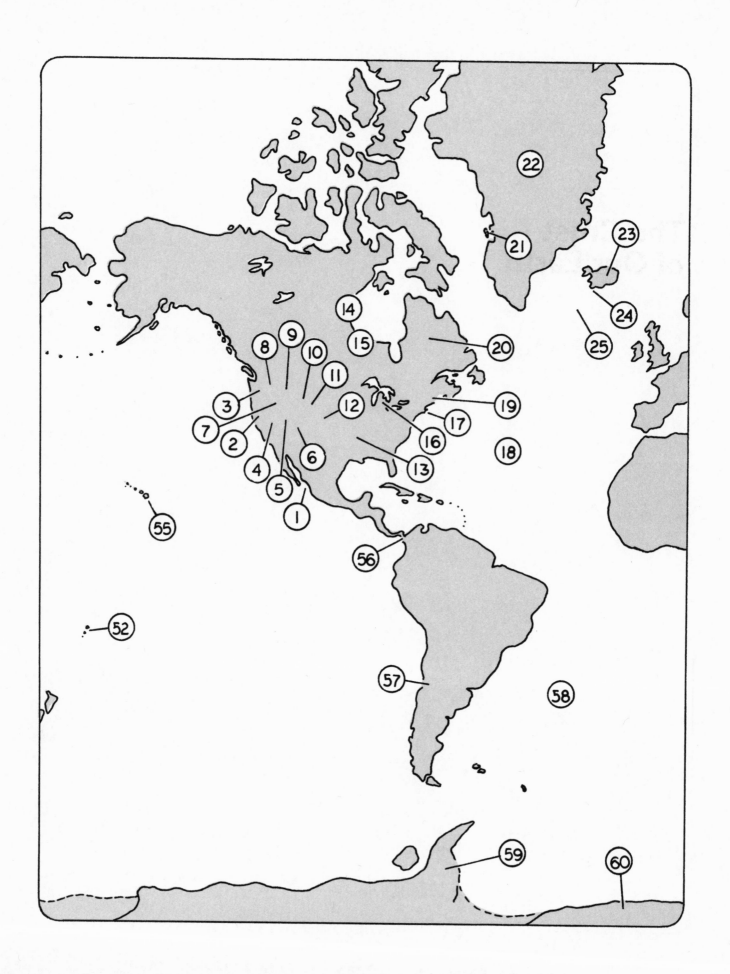

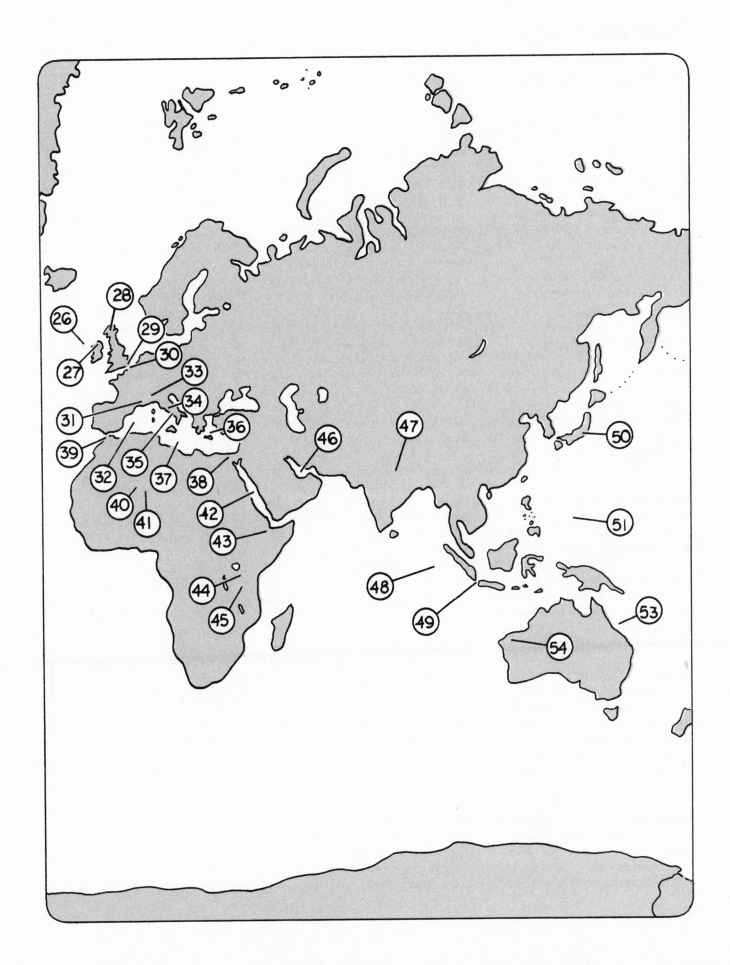

Introduction

A Revolution in Geology

A revolution has occurred in the earth sciences. It is hard to think of another revolution in the history of science that has been so quick in its execution or so sweeping in its effects. The textbooks of twenty years ago have been thrown out. The very foundations of geology, the bedrock principles of the science, have been overturned.

As we shall see, that last sentence is an apt metaphor. For at the heart of the new geology is the difference between a stable and a mobile crust of the earth.

The old geology was established in the late eighteenth and early nineteenth centuries by earth scientists such as James Hutton, William Smith and Charles Lyell. The great debate at that time was between the "catastrophists" and the "uniformitarians." The catastrophists believed the surface of the earth was shaped by unique moments of violence, a sequence of colossal deluges perhaps, such as the flood of Noah. It was a view that seemed especially compatible with the biblical story of creation. The uniformitarians rejected an earth history punctuated by calamity, and held instead that the crust was sculpted by imperceptibly slow processes of uplift, erosion, subsidence and deposition, acting over incalculable reaches of time. Hutton, Smith and Lyell compiled impressive observational evidence for the latter view. The earth, wrote Hutton at the end of his revolutionary book *Theory of the Earth*, shows "no vestige of a beginning, no prospect of an end." The debate was conclusively carried by the uniformitarians. Charles Lyell's *Principles of Geology*, published in 1830-33, laid the foundations of a uniformitarian science of geology.

Only twenty years ago, the uniformitarian view was still dominant in geology. Like Hutton and Lyell, most geologists believed the crust of the earth had always looked pretty much the way it does today, at least for the four billion years or so since the crust formed. Oh yes, some wrinkling of the crust had raised mountains, and weathering and erosion had worn them down. These mostly vertical movements were imagined to be not unlike the wrinkling that affects the peel of an orange as the fruit dries. Occasionally oceans invaded the interiors of sagging continents, allowing the deposition of sediments, and sometimes the continents lifted slightly and shrugged the waters back into their basins. Life evolved and had its own modifying influence on the atmosphere and crust. But these changes were slow, a working out over the eons of time of the same almost imperceptible changes that are occurring today. According to this older geology, the earth beneath our feet was "as solid as a rock" and could be relied on to stay that way.

We still accept many tenets of a uniformitarian geology, in particular the great age of the earth and the measured pace of geological process. But we also know that the long-term stability of the crust was an illusion, a trick of the incommensurability of the human and geologic time scales. The earth is not "as solid as a rock." Only a few tens of miles beneath our feet the rock is hot, plastic and in motion, a kind of motion known as convection. This internal turbulence breaks the crust and moves it about. The continents are dragged hither and yon, smashed together and riven apart, slowly growing in the process. The ocean floors are made and destroyed every few hundred million years. Earthquakes, volcanoes and mountain building are but the most obvious manifestations of this ongoing violence. According to the new geology, the motions which explain the major features of the earth's crust are horizontal, catastrophic and grand, not vertical, uniform and slight.

The idea of continents adrift on the crust of the earth had its genesis in a paper read to the Frankfort Geological Association in 1912 by a young German meteorologist named Alfred Wegener. Wegener had been struck—like others before him—by the remarkable jigsaw puzzle fit of the opposite coastlines of the Atlantic Ocean, especially by the way each bulge and indentation of Brazil had a corresponding concavity or convexity along the bight of Africa. Wegener argued that the continents had once been stitched together, as parts of a super landmass he called-*Pangaea* ("all-earth"). Then, said Wegener, several hundred million years ago Pangaea was ruptured and the continents drifted to their present positions, plowing like shallow rafts through the sea of rock that makes up the floors of the oceans.

Two years later, the world at war, Wegener was wounded during the German advance into Belgium. A long convalescence provided the opportunity to work out his ideas in detail. He presented these ideas to the scientific community in 1915, in a book called *The Origin of Continents and Oceans*. In this book Wegener amassed evidence from the fossil record and the structural record of the rocks to confirm the past unity of the continents and their subsequent drift.

Wegener's ideas had a skeptical reception. Geologists and geophysicists replied that the ocean floors were too rigid to permit the continents to barge through them. They pointed out that the roots of the continents were deep, more like icebergs than flat rafts. Besides, they asked, what conceivable force could push continents horizontally about the face of the planet? In 1926 the American Association of Petroleum Geologists organized a symposium of prominent earth scientists to definitively consider Wegener's hypothesis. The big guns of geology lined up and blasted Wegener's lightly armoured theory right out of the water.

And there the matter stood for thirty years. Then, during the late 1950s and early 1960s evidence derived from the magnetism of rocks seemed to confirm anew the drift of continents (see page 52). A new model was required, one that avoided the most solid objections to Wegener's theory. Recent studies of ocean floor topography contained many suggestive clues as to what the new model might be, for like a clever seamstress nature had concealed her stitching beneath a facing of water. In 1962, Harry Hess, building on ideas of Holmes, Dietz and others, gave form to the new model in a paper called *History of Ocean Basins*. The ideas proposed by Hess were radical, so antithetical to orthodox geology that he prefaced them as "geopoetry." According to Hess, the continents do not plow through the ocean basins like barges. Rather, the continents are passive passengers on a mobile crust that is continuously recirculated through the earth's hot interior. The key concept in Hess's essay was termed "sea-floor spreading" (by R. Dietz, 1961), for reasons we shall shortly see. Hess's ideas rapidly evolved into a comprehensive new theory of remarkable scope and power, the theory of plate tectonics.

The four essays that follow present the gist of the new theory and define some key concepts. A deeper explication of plate tectonics, and a fuller appreciation of the power of the theory, will unfold during our around-the-world geological journey. On that journey we shall stop at sixty fascinating sites on the earth's crust, and show how the new geology gives "rhyme and reason" to phenomena that were previously cloaked in mystery.

The supreme test of any good scientific theory is its economy and power—that theory is best which explains the most in terms of the least. By that standard, the theory of plate tectonics has proven marvelously successful. Within a decade of the publication of Hess's paper the revolution in geology was essentially complete, and few geologists continue to oppose the drift of continents. It is unfortunate that Alfred Wegener did not live to witness his idea confirmed. We now readily concede to the crust of the earth the dynamic aspect he envisioned in a bold and courageous leap of imagination.

Inside the Earth

The surface of our blue-green planet gives few hints of what's inside. The deepest man-made wells and mineshafts are only pinpricks on the earth's skin. Our present knowledge of the hidden interior is the product of a long process of induction and deduction, a "what-is-it" mystery painstakingly unraveled by geological detective work.

The Greeks were the first to measure the earth's diameter—by observations of celestial bodies—and came remarkably close to the present estimate of just under 8000 miles. By the early nineteenth century it became possible to use Newton's theory of gravitation to calculate the total mass of the earth and, since its size was known, its average density. The average density of the earth is 5.5 grams per cubic centimeter, or 5.5 times denser than water. But the average density of the rocks exposed at the surface is nearer to 3 grams per cubic centimeter. It is clear that some heavy material, denser than the surface rocks, must exist inside the planet to account for the discrepancy. Fortunately, nature gives us a hint of what that material might be. The meteorites that rain down upon the earth from space fall into two classes: stony meteorites with a composition not altogether unlike that of the earth's crust, and denser metallic meteorites consisting mostly of iron and nickel. If meteorites are typical of the stuff from which planets

formed, then it is not unlikely that iron and nickel might exist in quantity somewhere deep beneath our feet.

The structure of the earth's interior has been best revealed by the analysis of earthquake waves. When, somewhere in the crust, rock slips against rock the whole planet shudders. Seismographic stations around the world record these planetary spasms. A theoretical analysis of earthquake data has led to a fairly confident view of the earth's interior.

The drawing at right, showing the earth sliced open like a boiled egg, illustrates our current understanding. The yolk of the earth is the mostly iron-nickel core, with a radius about half that of the planet. The outer part of the core is molten, and fails to transmit those earthquake waves which require a solid medium for their propagation. Surprisingly, the inner core is not molten. The greater gravitational pressure at the greater depth raises the melting point and keeps the very center of the planet in a solid state.

The earth's mantle is the white of the egg and comprises the great bulk of the planet—82 percent of the volume and 68 percent of the mass. The material is stony, like the crust, although denser than crustal rocks. The massive inner part of the mantle, often called the mesosphere, is solid and readily transmits all kinds of earthquake waves. But a layer of

the upper mantle, where the pressure is low, is close to the melting point. This part of the mantle is called the asthenosphere, from the Greek word for "weak." The material of the asthenosphere is plastic and partly molten, and like iron beneath a blacksmith's hammer can flow under the influence of heat and gravity. As we shall see, the slow movement of rock in the asthenosphere has dramatic consequences for the earth's crust.

The cool, rigid skin of the earth is called the lithosphere, from the Greek for "stone." It is as thin, relatively speaking, as the shell of an egg. The drawing below, a profile of the crust across the United States, shows the lithosphere more or less in proportion to the curvature of the planet. An exact rendering would show wrinkles in the planet's skin, from the awesome summit of Mount Everest to the deep floor of the Pacific's Mariana Trench, occupying a layer only a few hundredths of an inch thick. Sketched in true proportions, the oceans and atmosphere would be films of water and air no thicker than a few pages of this book! Truly, the crust of the earth is as thin and smooth and hard as the shell of an egg. And the layers of air and water which are the rich and complex medium of our existence are as gossamer and insubstantial as the blush on a rose.

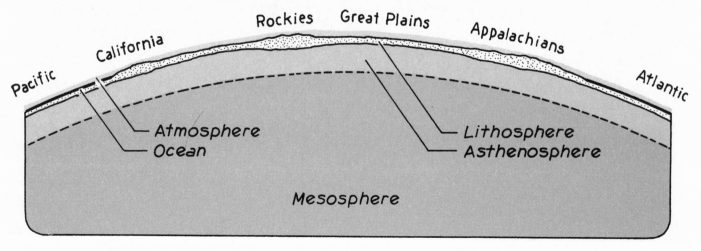

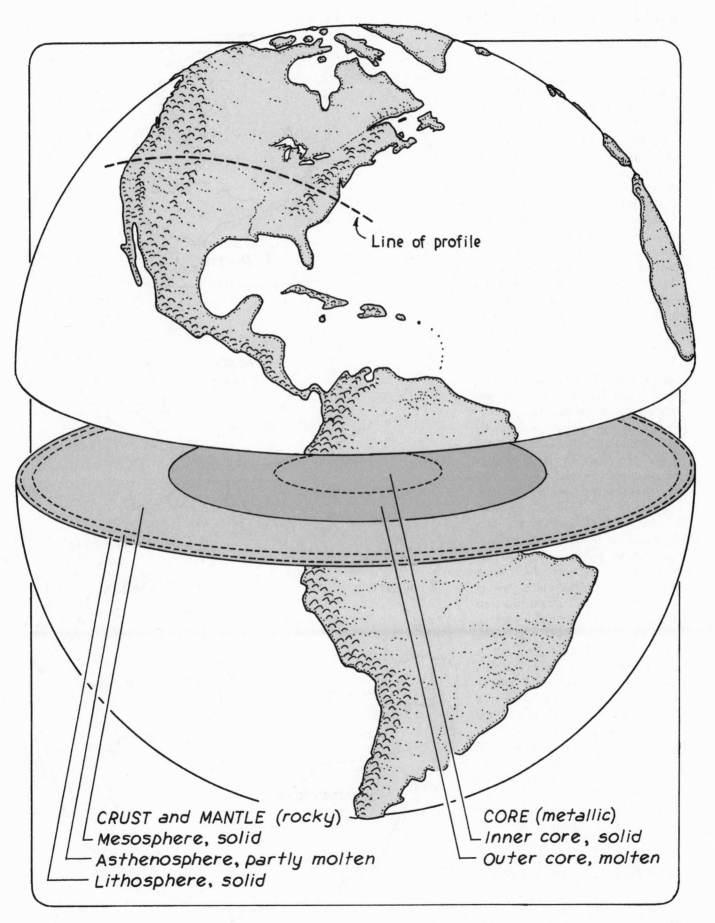

Line of profile

CRUST and MANTLE (rocky)
Mesosphere, solid
Asthenosphere, partly molten
Lithosphere, solid

CORE (metallic)
Inner core, solid
Outer core, molten

Crust of the Earth

In the new geology, the earth is more like an egg than a billiard ball. If the earth is an egg, with a metallic yolk and a rocky white, then the lithosphere is the shell. The lithosphere is rigid and typically about fifty miles thick. It rests on the weak, plastic, partially molten rocks of the asthenosphere. The continents with their deep roots are part of the lithosphere, and so are the floors of the oceans. Traditionally, only the very lightest rocks in the upper layer of the lithosphere are termed "crust." In the new geology, the structural boundary between rigid lithosphere and plastic asthenosphere appears more significant than the compositional discontinuity that has long distinguished the traditional "crust" and "mantle." In this book the term "crust" will be used more or less interchangeably with "lithosphere."

The continents have a distinctly different composition than the rocks of the ocean floor. The continents are blocks of light silicate materials with average properties similar to granite. Early in the earth's history, and continuing since, these light materials separated from the denser rocks of the mantle and now float on the upper mantle the way rafts of scum collect on the surface of a cooking soup. Because these "rafts" are less dense than the materials in which they float, they "ride high" like blocks of wood floating in water. It is this "riding high" or "bobbing up" of the continental rocks that creates areas of dry land above the level of the seas. The rocks of the continental crust are permanent surface features, although slowly growing in volume by the accretion of whatever light materials are differentiated from the mantle.

The floors of the oceans, on the other hand, are made of a denser rock called basalt, more closely resembling the materials of the mantle. We now realize that the rocks of

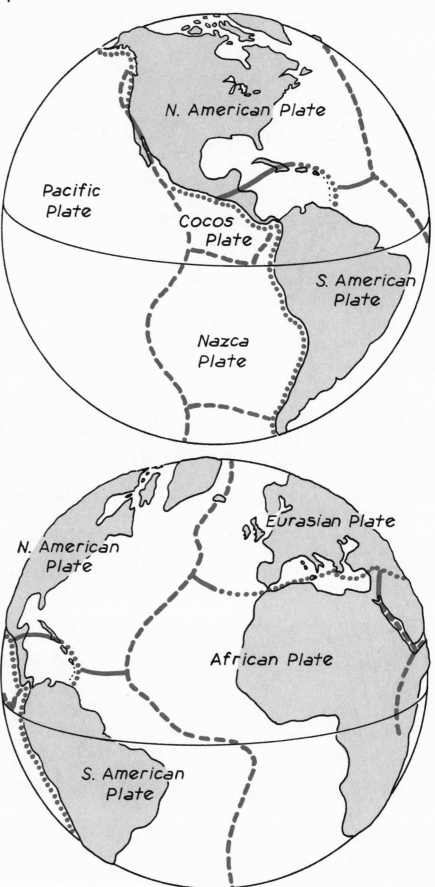

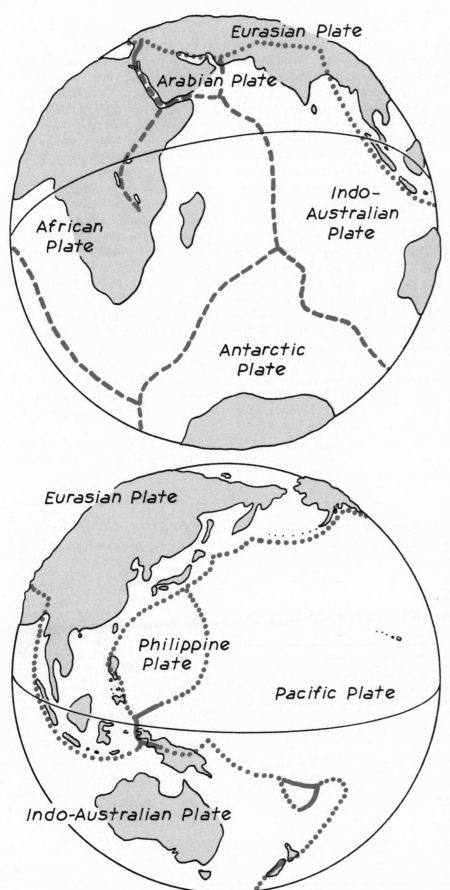

the ocean floors are continuously recycled through the mantle, rising at the mid-ocean ridges and sinking at the trenches (see page 9). No part of the present oceanic crust is more than a few hundred million years old.

Both ocean floor and continental rocks are part of the rigid "eggshell" of the planet. The eggshell is thin (in relation to the planet) and fragile. It is subject to stresses generated by the earth's internal heat and gravity. These stresses have cracked the eggshell into rigid sheets called plates. The drawings on these two pages show the major plates of the earth's surface. Some of these plates, such as the huge Pacific plate, consist entirely of oceanic crust. Most of the plates carry rafts of continental rock. The continents are embedded in the plates and share their motion.

These segments of "broken eggshell" do not constitute a stable, unchanging skin for the earth. Rather, they are dragged hither and yon about the surface of the earth by forces generated in the earth's interior. Along some plate boundaries— those corresponding to mid-ocean ridges—the plates move apart. Along these *divergent boundaries* (dashed on the maps) new oceanic plate is created by the solidification of mantle material rising from below. Along other plate boundaries—those corresponding to ocean trenches and young continental mountain ranges—the plates are pushed together. Along these *convergent boundaries* (dotted on the maps) oceanic crust is pushed back down into the mantle and continental rocks are squeezed upward. It is along plate boundaries that most geological activity occurs, including earthquakes, volcanoes and mountain building. Plate boundaries are "where the action is." Most of the stops on our around the world tour will be near present or past plate boundaries.

Plate Tectonics

The "cracked eggshell" plates of the earth's surface move! That was the astonishing discovery of the 1960s that launched a new appreciation for the old idea of continental drift, and for the first time offered an all-embracing theory of terrestrial geology. We shall see in the course of our travels how the drift of plates was discovered and confirmed. For the moment, a brief outline of the theory.

Beneath the eggshell crust of the earth, the hot, plastic material of the asthenosphere is in motion, a kind of motion called convection. Hotter, lighter material rises and cooler, heavier material sinks, just as the air in a room or pudding in a pan circulates vertically when heated from below. Along a worldwide system of mid-ocean ridges mantle material rises, lifting and fracturing the earth's crust. The crust is torn apart along the ridges (we shall see why in a moment), and lava (molten rock) oozes into the rift to fill the gap. As this material solidifies it forms new

ocean crust, a kind of rock called basalt. This new crust is itself soon fractured and pulled away in opposite directions with the growing, moving plates. The rate of motion is typically an inch or two per year. In this way, new ocean floor—new "eggshell" for the earth—is continuously created at the oceanic ridges.

But crust cannot be created in one place without being consumed in another. At convergent plate boundaries, one slab of cool, rigid crust is forced down beneath the other. When it is two oceanic plates which meet, the subduction of ocean floor can proceed either way. But when oceanic plate pushes against a continent, it is the ocean floor which is inevitably forced back down into the mantle. The lighter continental rocks resist subduction for the same reason a block of wood bobs back up when you try to push it under water. But under the force of convergence continents are crumpled, and great mountain ranges, such as the Alps or the Andes, are pushed up.

Earthquake and volcanic activity on the crust of the earth is almost entirely confined to plate boundaries. It is the bending, breaking and sliding together of the rigid crustal plates that cause most of the planet's earthquakes. Along the mid-ocean ridges these quakes are shallow, no deeper than the thickness of the lithosphere. The quakes can have a deeper focus on convergent boundaries, anywhere along the cool, rigid subducting plate. Along divergent boundaries, volcanoes are stoked by rising mantle material warmed by the earth's internal heat. Volcanic activity along convergent plate boundaries results from energy released in or along the diving (subducting) plate. The heat generated by subduction can have several causes: friction, compression, convection, and changes in mineral structure are a few of the sources of heat that have been proposed by geologists.

The nature of the force that drives the moving plates has been hotly

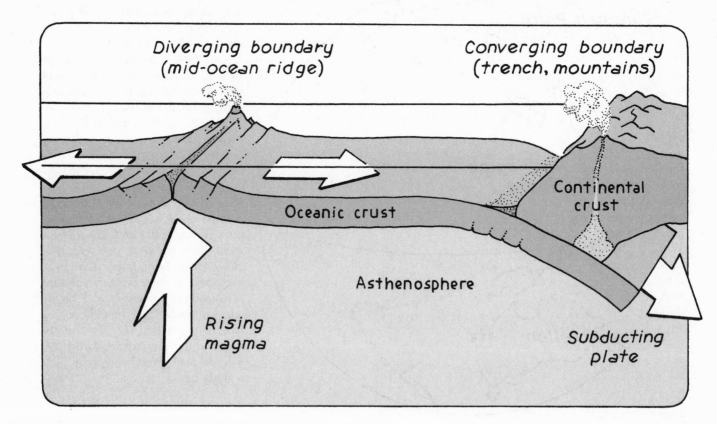

Diverging boundary
(mid-ocean ridge)

Converging boundary
(trench, mountains)

Continental crust

Oceanic crust

Asthenosphere

Rising magma

Subducting plate

debated by geologists. The four drawings at the right illustrate several of the driving mechanisms that have been suggested. 1) One of the most popular theories assumes the existence of convection loops in the asthenosphere. At some places hot mantle rock rises; in other places cooled, heavier material sinks. As the plastic rock of the asthenosphere flows beneath the rigid crustal plates, it exerts a frictional drag that pulls the plates along. 2) A second theory suggests that the force of rising mantle material along the mid-ocean ridges pushes the plates apart. 3) Try to pile up sand and gravity will pull it down. Push up the cool lithospheric plates at the oceanic ridges and gravity will pull them down. According to this theory, the lifted portion of the plates "slide down" the ridges and push the mostly horizontal plates forward. 4) Cool subducting plates are heavier than the surrounding mantle material. According to this fourth view, it is the weight of the subducting tongue of plate that drags the rest of the plate behind it.

In actual fact, all of the above forces may contribute to the drift of crustal plates. Although the mechanism of plate motion cannot be said to be fully understood, the reality of sea floor spreading and plate subduction is now almost universally accepted. Ocean floor is created at the mid-ocean ridges and consumed at the trenches. In this way, the basaltic rocks of the earth's crust are recycled through the mantle. No part of the earth's ocean floors appears to be more than 200 million years old. Whatever light materials are separated from the mantle by passing through the mill of plate creation and destruction are added to the ever-growing continents. As we travel around the globe we shall see how the dynamic engines of heat and gravity are always at work reshaping the crust of the earth.

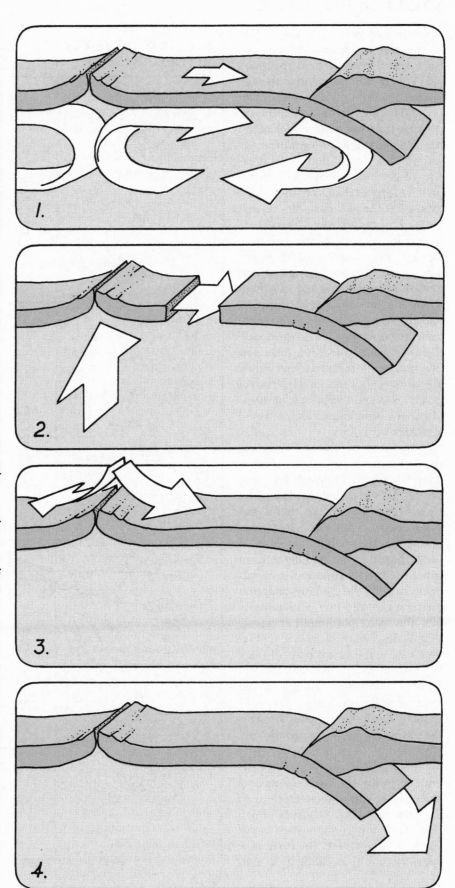

Geologic Time

Two hundred years ago most people in the western world believed the earth was five or six thousand years old. We are now certain that the age of the planet is at least 4.6 billion years! The idea of the immensity of geologic time was an invention of the late 18th and early 19th centuries. It was an idea closely tied to the creation of geologic science. A satisfactory interpretation of the rocks exposed at the earth's surface turned out to be possible only if geologists were allowed vast stretches of time for invisibly slow forces of uplift, subsidence, erosion and deposition to raise up mountains and tear them down. On the other hand, the story told by the rocks, once properly understood, confirmed on every side the great age of the earth. If the last two sentences sound circular, they are. In science, the engendering of grand ideas and the prosaic business of patient observation are entwined activities.

According to present theories, the earth condensed from a nebula of interstellar gas and dust at the same time as the sun and other planets. Its early history is uncertain, but a possible scenario follows. Sometime during the first half-billion years of earth history, the undifferentiated sphere of gravity-collected materials heated up to the melting temperature of nickel and iron. Radioactivity was the principal warming agent. When the heavy elements melted they sank to the earth's core, displacing lighter rocky materials and releasing gravitational energy in the form of still more heat. The consequent rise in temperature brought the entire planet, rocky mantle and metallic core, to or near to the melting point. The earth's first crust, together with the first oceans and atmosphere, formed shortly thereafter from the light elements which made their way to the surface of the partly molten planet. The earth now began to cool by radiating its heat to

MILLIONS OF YEARS BEFORE PRESENT	GEOLOGIC PERIODS	
2	CENOZOIC	Quartenary
		Tertiary
65	MESOZOIC	Cretaceous
135		Jurassic
210		Triassic
230	PALEOZOIC	Permian
		Carboniferous
		Devonian
		Silurian
		Ordovician
		Cambrian
575		Precambrian
4600		

Note: Scale not linear.

At least four ice ages occur
Appearance of humans

Beginning of the Andes

Grazing animals multiply
Grasslands spread
Collision between Africa and Europe,
   Beginning of the Alps
Collision of India and Asia, beginning
   of Himalayas

Mammals and birds flourish

Rocky Mountain uplift begins

Extinction of dinosaurs

Flowering plants
First mammals

Age of the dinosaurs
First birds
Opening of present Atlantic Ocean
Pangaea splits up

Cone-bearing trees
Reptiles established

Amphibians established
Collision of Europe and Asia, beginning
   of Urals
Collision of Africa and North America,
   Beginning of Appalachians
Extensive coal-forming forests
Winged insects

Plants and animals move onto land
First vertebrates, the fishes, appear

First animals with shells

Sexual reproduction
Respiration
Photosynthesis
Growth of continents
Origin of life
Crust forms; first oceans and atmosphere
Differentiation of core and mantle
Earth forms from solar nebula

---

space. By about 3.5 billion years ago, the earth had a solid crust physically similar to the crust today. At about the same time, life appeared on the planet and began its own work of modifying the crust.

A relative timetable of earth history was worked out by eighteenth- and nineteenth-century geologists who studied the stratified sedimentary rocks of the crust. The age of the strata relative to one another could be determined by their physical relationship (youngest strata on top, oldest on the bottom), and by the fossils they contained. Throughout the nineteenth century, geologists worked hand in hand with biologists to reconstruct the coupled evolution of the rocky crust and life. The names which geologists use to designate chapters of earth history (see chart) date from that time. The *Cretaceous* period, for example, takes its name from the Latin word for "chalk," and referred originally to chalk beds in France, Belgium and Holland that were studied in the early 1800s. Although geologists of the nineteenth century were able to consistently recognize rocks of, say, the Cretaceous period, they could only guess at the actual physical age of those formations.

The breakthrough in the creation of a quantitative timetable of terrestrial evolution came at the end of the nineteenth century with the discovery of radioactivity. The precisely measurable transformation rates of radioactive elements in certain igneous rocks can be used as a kind of natural clock to determine the time since those rocks were in the molten state. Radiometric dating made it possible to put absolute dates on the geologic time scale.

The earth's crust turned out to be very ancient indeed. On our armchair journey around the world we shall visit some of the oldest rocks of the crust, and also some of the youngest.

1 Tube Worms and Smokers

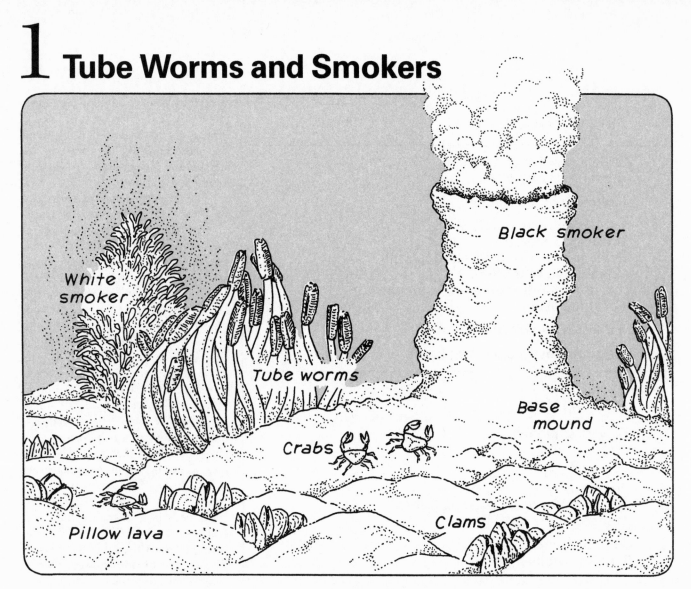

White smoker

Black smoker

Tube worms

Crabs

Base mound

Pillow lava

Clams

The crust of the earth is not uniformly active. Most of the North American continent, for example, is geologically quiet. East of the Rocky Mountains only an occasional deep tremor is likely to disturb the repose of the rocks. But along the western margin of the continent volcanoes belch fire and earthquakes jolt the landscape.

The western edge of North America lies along active boundaries of the "eggshell" plates that make up the earth's rigid crust. Central America and the northwestern United States abut convergent boundaries, where ocean floor is being thrust back down into the mantle [3]. Slicing across southern California the boundary between plates is known as the San Andreas Fault [2], and here the plates slide laterally against one another. Where the San Andreas Fault

passes into the Gulf of California it becomes a divergent boundary, a northern extension of the six-thousand-mile-long rift system known as the East Pacific Rise. Sea floor plates move apart along this submarine ridge and new ocean floor is added to the crust as magma (molten rock) rises from below to fill the rift. Here the North American continent has overridden the divergent plate boundary. Baja California is a sliver of continental rock the separating plates have torn from the flank of Mexico.

In recent years, while exploring the Pacific rift system with the deep-sea research vessel *Alvin*, scientists have made a surprising discovery. Flourishing at several locations on the otherwise almost barren sea floor are rich biological communities unlike any others on planet earth.

These unexpected oases of bizarre life forms are clustered around curious systems of hot water vents associated with the rifting crust. One such site, illustrated at right, is located near the mouth of the Gulf of California.

Researchers who have studied these remarkable sites speculate that cool sea water percolates through fissures in the sea floor and is heated near chambers of magma that lie just below the rifting crust. The heated water is then expelled upward through vents above the magma chambers. The expelled water is rich in dissolved minerals—including oil and hydrocarbons—which have been steeped from sea-floor sediments and fresh rock the way coffee is brewed in a percolator. Some of the dissolved minerals precipitate out of the water and build up exotic chimneys above the vents. One *Alvin* passenger described the chimneys as "Japanese pagodas." Some of the chimneys spew out water blackened by sulfide particles and have been dubbed "black smokers." Others gently emit water that is milky white and are called "white smokers." The chimneys sit on heaps of precipitated minerals and are covered with mats of yellow-orange bacteria. The sea floor surrounding the strange towers consists of pillowlike mounds of fresh black volcanic rock.

Most life forms on earth are part of food chains that ultimately depend for energy on the photosynthesis of sunlight—plants trap the sun's light, animals eat plants. But the creatures of the hot, lightless world of the East Pacific Rise apparently rely for sustenance on a flow of energy directly from the earth's interior. Most curious of all the creatures of the vent colonies are the giant tube worms, some ten feet long, that cluster about the smokers, waving bright red plumes that extend from protective sheaths. The worms have no mouth or gut and seem to obtain their nourishment from the nutrient-rich sea water expelled by the vents. This strange harvest is accomplished through a remarkable collaboration with bacteria that inhabit the long

bodies of the worms. The worms are not alone in their dark bower. Huge clams live in the crevices between the pillows of lava. White crabs the size of dinner plates scamper about the chimneys.

It would be hard to imagine an environment more hostile to life than the crest of the rifts a mile or more beneath the sea, a world of extremes of heat and cold and perpetual darkness. The communities of creatures that survive in this lightless world affirm the astonishing resourcefulness of life on the crust of the earth.

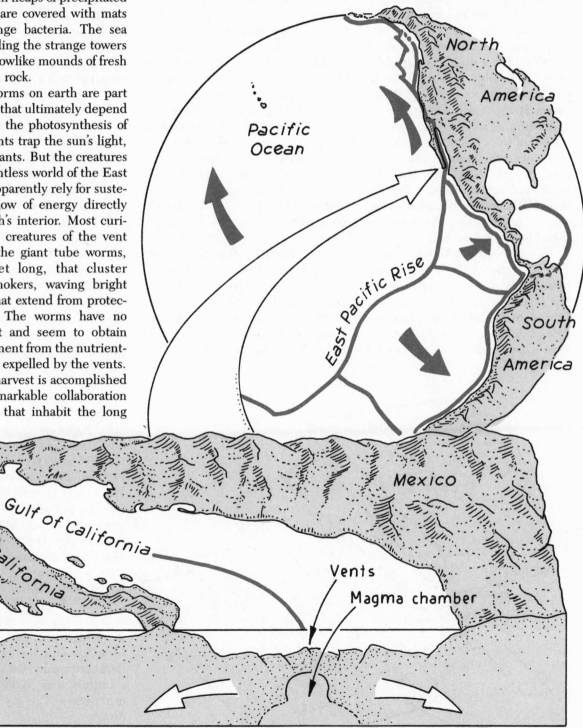

2 The San Francisco Quake

Early on the morning of April 18, 1906, the earthquake that would devastate the young metropolis of San Francisco had its beginning at a point on the San Andreas Fault some thirty miles northwest of the city. Quickly the shock ripped southward along the fault, diving several times beneath the sea, finally tearing through the countryside just southwest of the city like the crack of a whip. One eyewitness described it like this: "There was a deep rumble, deep and terrible, and then I could see it actually coming up Washington Street. The whole street was undulating. It was as if the waves of the ocean were coming towards me, billowing as they came." Seven hundred people died that day. The quake and the subsequent fire left tens of thousands homeless. Four square miles of the city were reduced to charred rubble.

The 1906 earthquake was inevitable, as is the earthquake that will someday in the future again wrack San Francisco. The city has the misfortune to lie astride the San Andreas Fault, one of the longest and most active breaks in the earth's solid crust. The map immediately to the right shows the city of 1906 and its proximity to the lively fault. The fault is part of the long system of rifts and faults that marks the boundary between the Pacific plate and the North American plate. As the map at far right shows, the San Andreas Fault is an extension of the rift that is presently opening up the Gulf of California and tearing Baja California away from the Mexican mainland.

Along the fault's tortured course through California the plates are not pulling apart, as in the gulf to the south, nor squeezing together, as they do along the coasts of Oregon and Washington. Rather, they are sliding against one another, the Pacific plate moving slowly but inexorably toward the northwest. The southwestern edge of California, along with all of Baja California, is a sliver of continental crust that has become affixed to the Pacific plate and rides that plate northward. North of Los Angeles, where the sliver of continent grinds against the deeply rooted Sierra Nevada and San Bernadino Mountains, it is fractured and deflected to the west.

The average motion of the plates along the San Andreas fault is several inches per year, sufficient to carry Los Angeles abreast of San Francisco in 10 or 20 million years. The motion is not smooth, but proceeds in fits and starts (see illustration below left). As the plates try to slide against one another, friction prevents a steady displacement (1). Stress builds up as the weakened rocks near the fault distort (2). When the accumulated strain becomes too great the rock breaks at the fault and rebounds elastically (3). Once initiated, the break can propogate along the fault—moving at speeds of thousands of miles per hour—for great distances. The 1906 earthquake affected several hundred miles of the fault, and horizontal displacements on some sections were as great as twenty feet.

The hilly shore of San Francisco Bay is one of the most densely populated regions in the world. It is home for nearly five million people. The modern city of San Francisco is far more vulnerable to earthquakes than the city of 1906. Much of the metropolitan area, including the airports, has been built on the unsteady foundation of filled-in land reclaimed from the bay. Great bridges, skyscrapers, dams and reservoirs, elevated highways and an underground rapid transit system are particularly at risk. The 1906 quake caused about $400 million damage, worth about $1.6 billion in today's dollars. It has been estimated that the same quake would do $30 billion damage to the modern city. Deaths might number in the tens of thousands.

It is certain that another major quake will someday lash the San Francisco region; the only question is when. Some geologists believe that strain along the fault has already been rebuilt to the level prior to the 1906 earthquake. Others believe that a century may pass before the city experiences a replay of the earlier disaster.

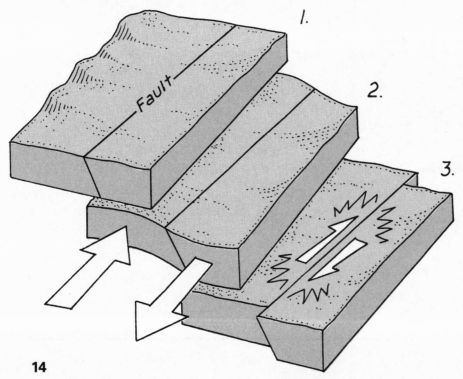

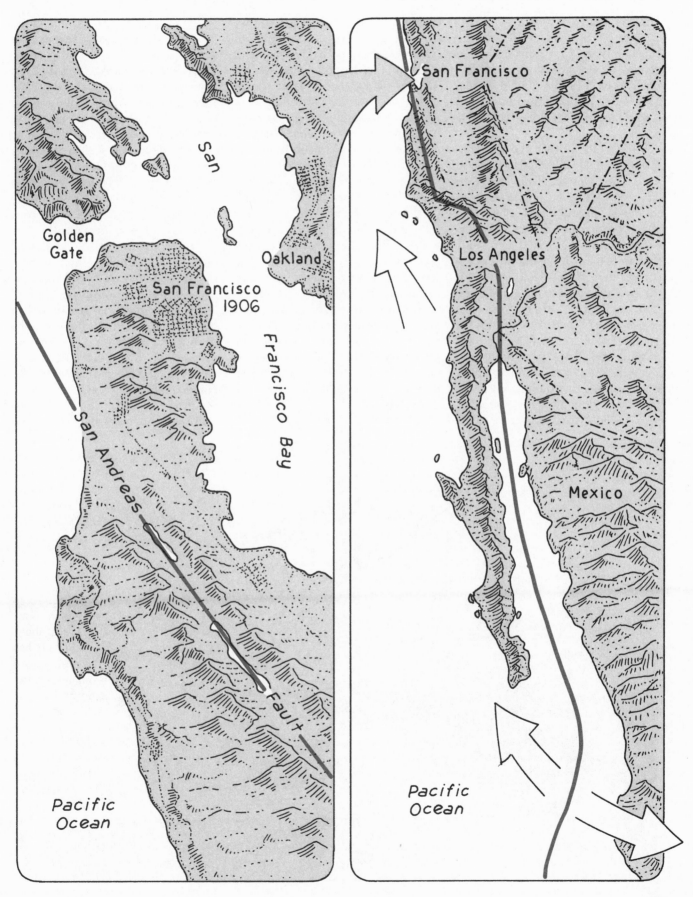

Golden
Gate

San

San Francisco
1906

Oakland

Francisco Bay

San Andreas

Fault

Pacific
Ocean

San Francisco

Los Angeles

Mexico

Pacific
Ocean

15

3 The Fires of Mt. St. Helens

The spectacular eruption of Mt. St. Helens on May 18, 1980 was the latest episode in a long history of geological violence along the western seaboard of the United States. The source of that unrest lies in the continuing collision of lithospheric plates.

Fifty million years ago the East Pacific Rise [1] still lay offshore along the entire western flank of North America. Then as now, the ocean floor was pulled apart along this great submarine ridge, and new ocean crust was created from rock that welled up from the slushy asthenosphere to fill the breach. But the ocean floor could not grow in one place without being consumed elsewhere. That part of the Pacific sea floor to the east of the Rise was pushed back down into the mantle along the margin of the North American continent, creating an ocean trench and crumpling the continent. Fifty million years ago, the western

coastline of the United States may have been similar to the present Pacific coast of South America, with a range of high mountains running parallel to a deep offshore trench [51].

About 30 million years ago the North American continent began to approach the East Pacific Rise. The first part of the continent to override the axis of sea floor spreading was the coast of southern California and northwestern Mexico. As spreading axis and subduction trench converged, the intervening oceanic plate was totally consumed. The deep sediments in the trench were caught in the squeeze and heaved up to form the coast ranges of California. These rocks became affixed to the Pacific plate and today ride the plate northward. Where the spreading axis hit the Mexican coast end-on it ripped Baja California from the mainland.

Off the present coasts of Oregon

and Washington, a small piece of the old East Pacific plate remains, not yet fully destroyed by subduction. This sliver of "broken eggshell" is known as the Juan de Fuca Plate (sometimes the Gorda Plate) and the remaining northern stretch of the East Pacific rift system is known as the Juan de Fuca Rise. The tiny Juan de Fuca Plate grows at the rise and is almost immediately consumed beneath the continent. As the fifty-mile thick crustal plate is forced back down into the upper mantle it is heated by friction, compression, and other factors. This heat melts parts of the descending plate and adjacent lithosphere, forming pockets of magma. The magma, less dense than the surrounding rock, literally melts its way to the surface. It was the pressure of this rising magma that blew away the top and side of Mt. St. Helens. The drawing below shows the extent to which Mt. St. Helens "blew its top." Nearby towns were

Mt. St. Helens
(----- before eruption)

smothered by several inches of ash. High altitude winds carried airborne ash all the way across the continent in three days, and around the world in seventeen days.

Mt. St. Helens is one of more than a dozen volcanic peaks in the Cascade Range of the American northwest. The earliest volcanic activity in this part of the country goes back hundreds of millions of years, but most of the range as we see it today was created relatively recently. Only 4600 years ago Mt. Mazama blew its top in the colossal explosion that created Oregon's Crater Lake. That blast expelled forty times more ash and pumice than the 1980 eruption of Mt. St. Helens. Since the Mt. Mazama catastrophe, Mt. St. Helens has been the most active peak of the range.

The trench where the Juan de Fuca plate slides beneath the continent is filled with sediments and few earthquakes occur along the diving plate. This suggests that subduction of the plate may be temporarily stalled. But volcanic activity in the American northwest will not cease until the continent stops pushing against the floor of the Pacific. Even the complete destruction of the Juan de Fuca plate will leave the area with an uncertain future. Mt. St. Helens and its sister peaks in the Cascades will erupt again.

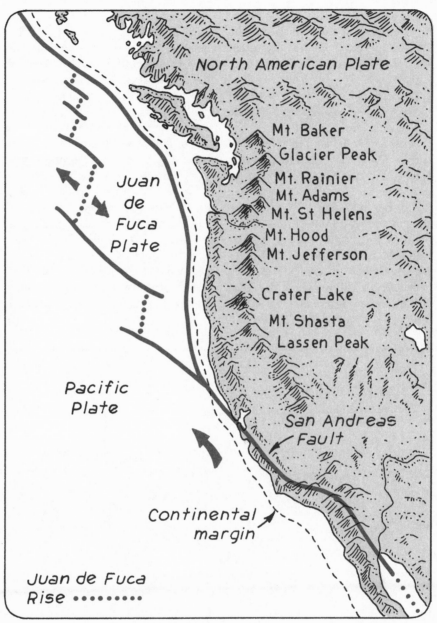

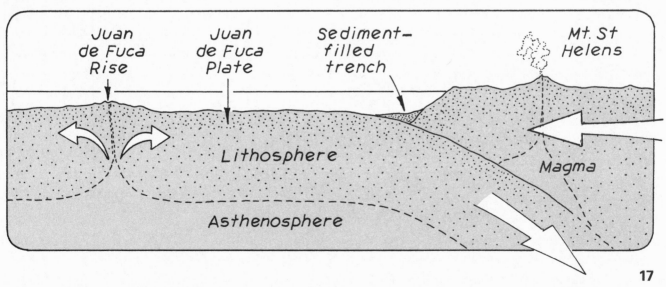

4 Highest and Lowest

A journey of no more than one hundred miles will take you from the highest elevation in the forty-eight contiguous United States to the lowest.

Mount Whitney soars 14,494 feet above sea level, rising with untypical abruptness from the floor of the Owens Valley in southeastern California. Only a few mountain ranges away to the east is the bleak depression known as Death Valley, dipping to 282 feet below sea level. Nowhere else on the continent is the vertical relief of the earth's surface so dramatic. Mount Whitney and the other lofty peaks of the Sierra Nevada help make Death Valley one of the most forbidding environments on earth. Moist prevailing winds from the Pa-cific are pushed up by the mountains. As they rise they cool, and the moisture they carry is precipitated as rain or snow on the western slopes of the Sierra. It is dry air that reaches the valleys to the east. Thirsty oaks, giant sequoias, lodgepole pines, firs and hemlocks march up the western side of the range. Down the eastern flank, pinon pines and junipers give way to the desert-loving sagebrush.

The broken topography between Mt. Whitney and Death Valley is part of what geologists call the Basin and Range Province. The province includes much of southeastern California and almost all of Nevada. It is characterized by ragged north-south tending mountain ranges separated by low, flat desert valleys.

Forty million years ago the Basin and Range Province was the site of the kind of explosive volcanic activity we now associate with the Pacific northwest [3]. Volcanoes similiar to Mt. St. Helens belched ash and lava. At that time the East Pacific Rise still lay offshore along the entire California coast, and there was a subduction trench at the coast itself where the East Pacific sea floor was driven beneath the continent. The coast was at that time somewhere near the left side of the drawing below, where the San Joaquin Valley is today.

But the continent was on the move toward the west. At last it pushed up against the East Pacific Rise, and the spreading axis and the subduction zone came into contact.

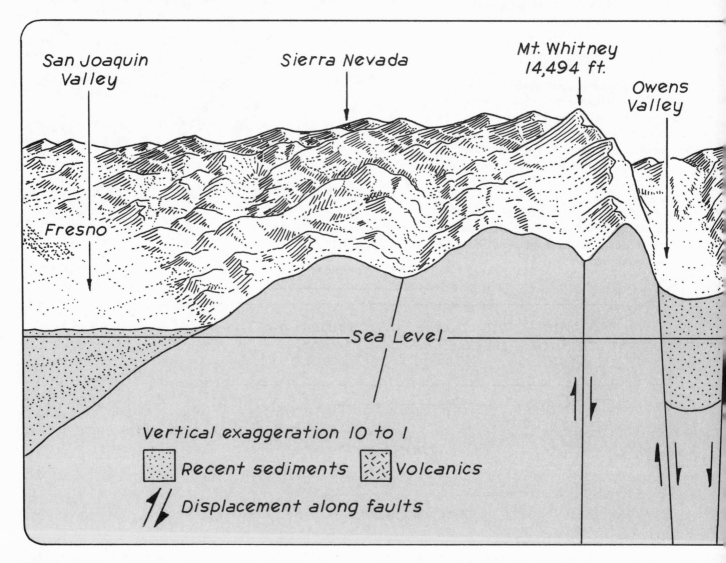

San Joaquin Valley · Sierra Nevada · Mt. Whitney 14,494 ft. · Owens Valley · Fresno · Sea Level

Vertical exaggeration 10 to 1

▫ Recent sediments   ▨ Volcanics

⚡ Displacement along faults

The volcanic activity in the Basin and Range Province changed character. Viscous, explosive lavas gave way to more liquid basaltic lavas that welled up from deep fissures and spread out in great sheets over what was still relatively flat terrain.

Beginning at roughly the same time, the entire crust between the Sierra Nevada and the Rockies was pushed up in a great barrel arch. The coincident tension in the crust and relaxation in east-west pressure caused the crust to break apart along north-south tending lines. Blocks of the crust dropped along these fault-lines to form basins. The basins immediately began to fill with sediments eroded from the adjacent ranges. To the west, a huge slab of crustal granite was tilted up along one of the faults to form the high Sierra.

The crust of the earth beneath the Basin and Range Province is only twenty miles thick, compared to twenty five miles beneath the Sierra Nevada and the Colorado Plateau to either side. There is an unusually high flow of heat up through this thinned crust. The nature of the forces that have stretched and fractured the crust of the Basin and Range Province, and which generate the heat that flows from below, are hotly debated by geologists. Most geologists agree that the spectacular vertical geography of Mt. Whitney and Death Valley are somehow related to the clash of lithospheric plates at the continent's edge. The dragging force exerted on California by the motion of the Pacific plate along the San Andreas Fault may be pulling the crust apart. Remnants of old subducted sea floor plate, cut off when the subducting trench and the East Pacific spreading axis came into contact, may still exist below the Basin and Range Province as a partially melted upward-spreading mass of lighter rocks exerting pressure on the crust above. It has even been suggested that the overridden East Pacific spreading axis may still exist beneath Nevada, threatening to tear the American west asunder and separate California from the eastern states with a new arm of the sea.

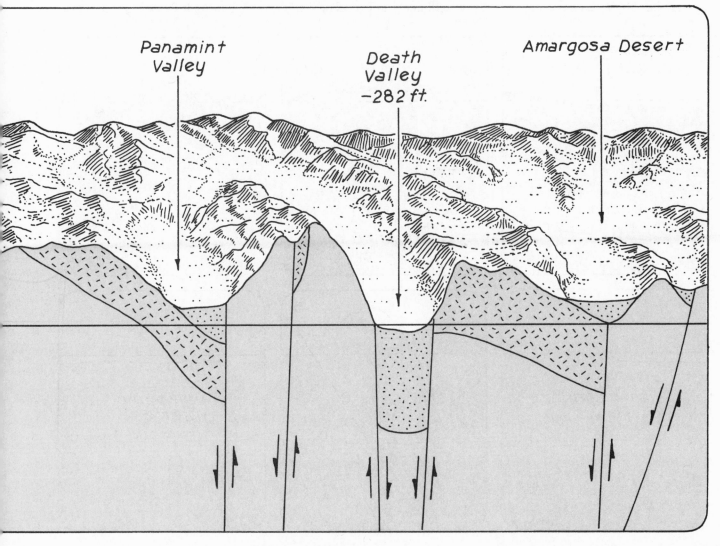

5 The Abyss of Time

Geologists have only recently begun to understand how the topography of the American west is related to the clash of plates along the continental margin. A rough recounting of some events of the past 50 million years has been given on the first few pages of this book. But the earthquakes on the San Andreas Fault, the fires of Mt. St. Helens, and the jagged geography of the Basin and Range Province represent only the last few pages of the geological history of the region.

Nowhere else does the crust of the earth provide a more tantalizing glimpse into the "dark backward and abyss of time" than on the walls of that great crustal incision, the Grand Canyon. The record of the rocks on the canyon walls takes us back through the earth's history halfway to the planet's beginning 4.6 billion years ago. The story is reconstructed in the drawings below.

The Grand Canyon is the work of the Colorado River and offers a startling example of the erosional power of moving water. The deepest part of the canyon slices a mile into the earth's crust. At the bottom of the gorge the river has exposed the 2-billion-year-old Vishnu schists. These ancient metamorphic rocks were created when volcanic lavas and marine sediments were subjected to conditions of extreme heat and pressure that altered their mineral composition and structure. By identifying the parent rocks, geologists can surmise that billions of years ago there was a subsidence of the crust coupled with the deposition of sediments and occasional vol-canic activity (1). The total thickness of the sediments might have been as great as five miles. At that time the region may have lay at the margin of an earlier continent. It possibly resembled the Gulf of Mexico of today.

About 1.7 billion years ago the region was caught in a squeeze that forced up high mountains and metamorphosed the deeply buried lavas and sediments, creating the Vishnu schists (2). There were simultaneous intrusions of molten rock that formed the granites exposed deep in the gorge today. It is not hard to imagine continents colliding in the record of those colossal events.

Then erosion went to work, cutting down the high peaks. As the burden of the mountains was removed the crust rose, ultimately exposing the schists and granites that

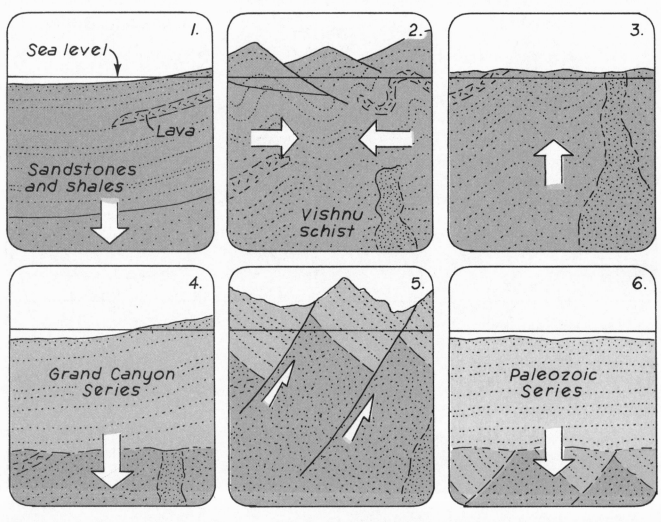

had been formed far below the surface (3).

The relatively level platform created by erosion now subsided, and new sediments (the Grand Canyon Series) were deposited on the descending base (4). The earliest of these strata contain the first local record of life, fossils of algae that thrived in shallow-water seas a billion years ago.

Again the rocks tell the story of a great deformation. The ancient metamorphic base and the more recent sedimentary rocks were heaved up and broken by faults (5). Great fault block mountains, such as are now common in the American west, stood high above the landscape. And again the forces of erosion attacked the mountains, eventually cutting them down to a sea-level plain. Only downfaulted wedges of the sedimentary strata remained.

Beginning about 570 million years ago there was a further episode of subsidence and deposition (6). The strata from this period, laid down alternately on land and beneath shallow seas, consist of shales, sandstones and limestones. They contain a vivid fossil record of the evolution of life throughout the Paleozoic Era, and include trilobites, fish and ferns. The arrival of the reptiles is recorded in fossil footprints in the higher formations. The car parks, campgrounds and hotels on the rim of the canyon bring the story of life on earth right up to date.

At about the time North America overran the East Pacific Rise (30 million years ago), the area of the Grand Canyon began to rise, straight up like the floor of an elevator. The Colorado River, or an ancestor of that river, began cutting down into the Paleozoic sediments, then into the Grand Canyon Series, at last exposing the Vishnu schists. This incision is the canyon of today (7), with its spectacular story of crustal dynamics exposed for geologists and tourists to read like a book. The causes of the events recorded on the walls of the Grand Canyon are not known with certainty, but 2 billion years of push and shove of crustal plates are much in evidence.

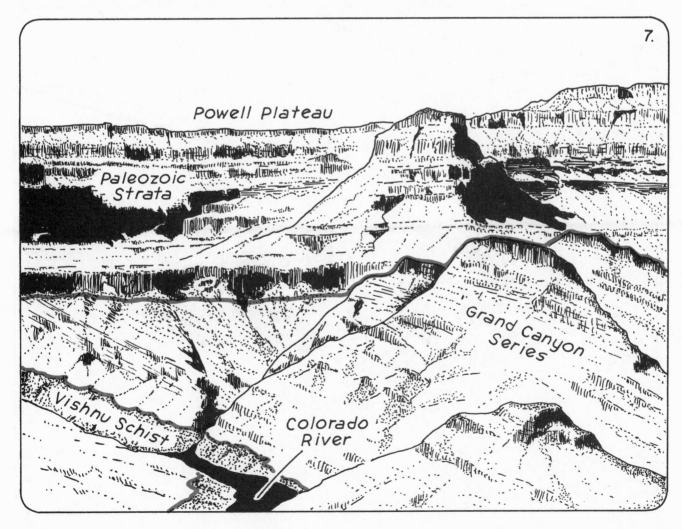

7.

Powell Plateau

Paleozoic Strata

Grand Canyon Series

Vishnu Schist

Colorado River

6 Meteor Impact!

Early in the history of the solar system, when space was still cluttered with the materials of its formation, the planets and their moons were heavily bombarded by meteorites. Some of the members of the solar system (Mars, Mercury and our moon, for example) still show the scars of that primordial rain of iron and stone. On our dynamic planet earth, erosion by weather, water and ice and the continuous reshuffling of crustal plates have erased most of the evidence of that early cratering. But the solar system is not yet completely clear of the cosmic debris which was left over at its birth. Small meteorites continue to rain down upon the planets, and occasionally the earth is struck by an object large enough to excavate a sizable hole.

Dozens of meteorite impact craters have been recognized on the crust of the earth. In most cases, erosion has removed all but the shattered root zones of the craters.

The most famous terrestrial impact crater is in the desert near Winslow, Arizona. The crater is three-quarters of a mile in diameter and 600 feet deep. The rim rises more than 100 feet above the surface of the surrounding limestone plateau. The feature is known today simply as Meteor Crater, but it was not until 1891 that its origin was recognized as cosmic rather than volcanic. There are, in fact, ancient volcanic craters in the area that helped to confuse the issue.

Toward the end of the last century many fragments from an iron-nickel meteorite were found on the surface around the crater and buried within the debris that partly fills the crater's bowl. This led David Moreau Barringer, a prominent industrialist, to suppose that an economically valuable mass of meteoric iron, perhaps weighing millions of tons, might underlie the crater. Barringer bought the crater and surrounding land from the US government, but extensive drilling failed to reveal the hoped-for treasure.

Barringer did not realize that the meteorite that excavated the crater was largely fragmented and dispersed upon impact. Still, over twenty five tons of fragments have been recovered from the area, giving a hint as to the nature of the object that smashed into Arizona. The ob-

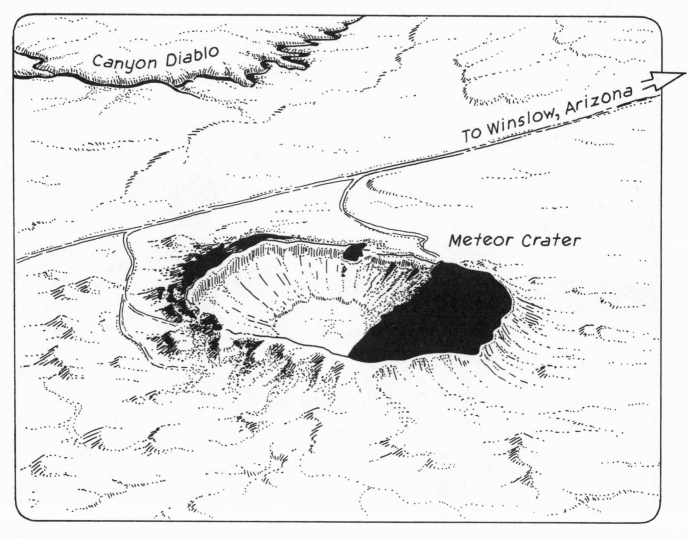

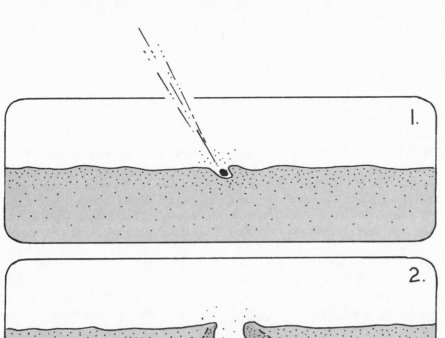

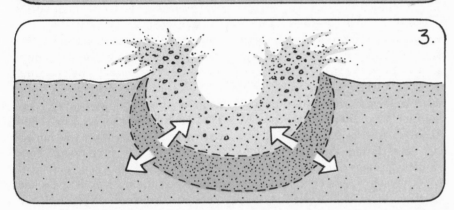

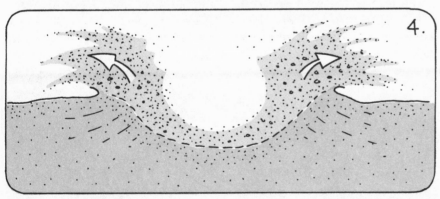

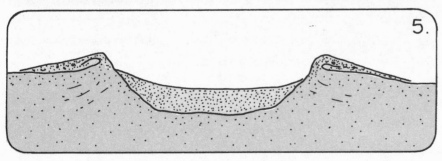

ject may have been 100 feet in diameter and weighed 100,000 tons. It was probably traveling through space at tens of thousands of miles per hour when it collided with earth.

The drawings at left reconstruct the excavation of the crater. The impact of the meteorite (1) set up a compressional shock wave (2) that moved outward from the impact point at supersonic velocity. The high pressure wave crushed the rocks of the horizontal strata, altered minerals to a glasslike composition, and fused fragments together. A reflected shock wave (3) bounced back from the compressed earth blasting the shattered and altered rock from the crater (4). The uppermost strata were folded back at the lip of the crater and some of the debris from the blast fell back to partially fill the excavated bowl (5).

The impact that excavated Meteor Crater seems to have occurred about 25,000 years ago, only yesterday on the geological time scale. Since that time the rim of the crater has been reduced by erosion and some of that eroded material has been washed down to cover impact debris and meteoric materials in the bottom of the bowl.

The young and relatively undisturbed Arizona crater offers geologists an unparalleled opportunity to study the cratering process. It also provides an ominous reminder that meteorite impacts remain a fact of life for planets in our solar system. If an object the size of the meteorite that blasted the Arizona crater were to impact today in the middle of an ocean, the Atlantic for example, the result would be catastrophic. Tidal waves would devastate coastal regions on both sides of the ocean. Great cities such as London, Rio and New York would be deluged. The loss of life and property would be too terrible to contemplate.

7 The Vanished Lakes

What geologists call the Basin and Range Province [4] roughly coincides in its northern portions with the geographic province known as the Great Basin. The Great Basin is hemmed in on the west by the Sierra Nevada and on the east by the Rockies. It has no outlet to the sea. The prevailing winds in the Great Basin are from the west. Warm, moist air from the Pacific is forced upward as it crosses the Sierra. At the higher altitudes it cools and the moisture it carries is precipitated as rain or snow on the western slopes of the mountains. It is air wrung dry of moisture that reaches the Basin. What little water falls in the Basin as rain or snow, mostly in the winter months, evaporates on the broad, flat desert floors. Nevada is called the Sagebrush State; the low clumpy shrub has a tolerance for arid, alkaline soils. Along the rare watercourses, cottonwoods and willows eke out a sparse existence. In the upland ranges, pinon pines and junipers struggle to hold their own.

But the Great Basin has not always been so arid. Many of the dry, closed depressions of the Basin were once filled with water. Owens Valley, Panamint Valley and Death Valley (see drawing pages 18-19) were once a string of interconnected lakes. The two largest of the ancient lakes of the Great Basin, Lake Lahontan and Lake Bonneville, are shown in color on the map at right. Dozens of smaller, now vanished lakes are not shown. The Great Salt Lake is all that remains of Lake Bonneville, and Pyramid Lake in Nevada is one of the last briny remnants of Lake Lahontan. The extent of these ancient lakes can be traced in wave-cut or wave-built beach terraces that were left high and dry on the sides of ridges when the water level dropped. The drawing below, adapted from an old Geological Survey illustration, shows several wave-shaped terraces near Logan, Utah, representing successive levels of Lake Bonneville.

There seems to have been several periods within the last tens of thousands of years when water accumulated in these basins. The rise and fall of the lakes were undoubtedly linked to the advances and retreats of the great ice sheets that covered much of the northern part of the continent during those times [14]. Climatic changes during the ice ages sometimes brought cooler, wetter weather to mid-latitude deserts worldwide, including those of the Great Basin. The broken, down-faulted valleys of the Great Basin provided ready receptacles for this moisture.

Lake Bonneville once covered 20,000 square miles of northwestern Utah, an area about equal to that of present Lake Michigan. It reached a maximum depth of nearly 1000 feet, compared to the 30-foot depth of the Great Salt Lake today. At some point in time, the surface of the slowly-filling landlocked lake reached the elevation of Red Rock Pass in southeastern Idaho. Here Lake Bonneville at last found an outlet to the sea, down the valleys of the Snake and Columbia Rivers. Spilling over the floor of the pass, the waters of the lake rapidly excavated a deep channel in the unconsolidated soil. There was catastrophic flooding along the Snake River as the lake poured out its contents. Huge boulders strewn on flat surfaces high above the present day river channel provide striking evidence of the force of this great ice age deluge. The boulders seem to have been ripped from valley walls and hurled about like so many grains of sand. Valleys were stripped and deepened. Only when the floor of the pass had been scoured down to bedrock did the mighty outpouring of water cease. The Snake River flood was prodigious, and may have been wit-

Wave-cut delta terraces of former Lake Bonneville

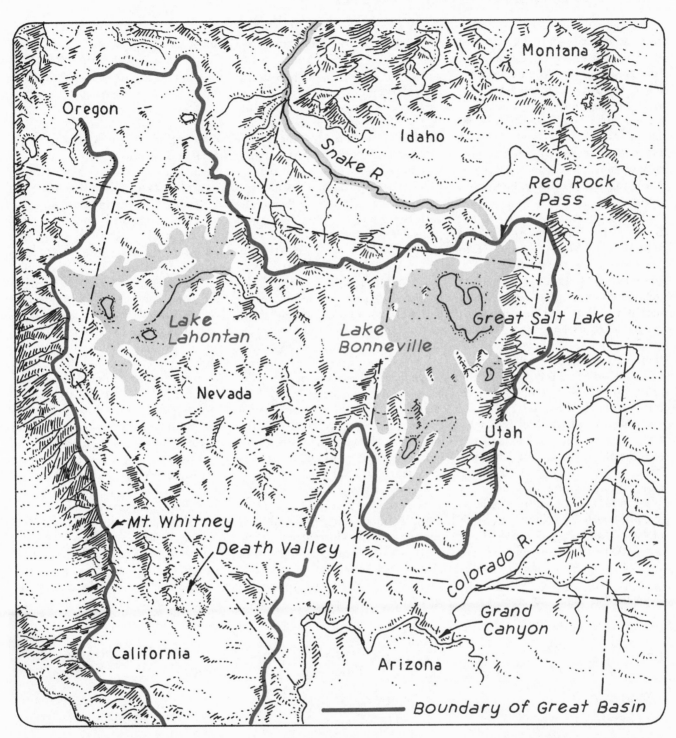

Boundary of Great Basin

nessed by prehistoric men and women who were among the earliest migrants to the North American continent. But the outpouring of Lake Bonneville into the valley of the Snake was not, as we shall see, the greatest flood that has occurred on the crust of the earth [8, 39].

Lake Bonneville stood for a long time at the level determined by the bedrock sill of the Red Rock Pass. Large deltas were built up at the mouths of rivers flowing into the lake from the mountains to the east. Salt Lake City sits on one of these ancient delta platforms.

As the last glacial epoch came to an end, precipitation diminished in the Great Basin and evaporation increased. Most of the ice age lakes dried up completely. Lakes Lahontan and Bonneville shrank to salty vestiges of their former selves. Today, at Bonneville Flats in western Utah, automobiles race on the smooth, caked salt of the old lake floor.

8 The Spokane Flood

The great outpouring of the waters of Lake Bonneville into the valley of the Snake River was not the only, nor the greatest, catastrophic flood to affect the American northwest. At least once and possibly several times during the last ice age eastern Washington state was subjected to a scouring deluge of almost unimaginable proportions. The affected area is known today as the Channelled Scablands.

The Scablands are part of the basalt plateau that covers much of southeastern Washington, eastern Oregon and southern Idaho. The plateau consists of almost undeformed layers of lava that seem to have poured out onto the crust of the earth during Miocene times (10 to 20 million years ago). The lava flows are thousands of feet thick and appear to have welled up from beneath the crust through deep fissures. There is little evidence of the kind of explosive volcanic activity that is characteristic of Mt. St. Helens and the other volcanoes of the Cascade Range to the west, all of which are more recent than the upwelling that formed the basalt plateau. The lavas of the plateau rose from below like molasses, and spread out to bury over 200,000 square miles of the Pacific northwest under layer after layer of dense black rock. In this dramatic "overturn" of the earth's upper layers, more rock than might be found in a large mountain range was removed from below the crust and added on top. The episode was undoubtedly related in some not yet well understood way to the subduction of the Juan de Fuca plate beneath the western margin of the continent [3]. An east-west cross-section of the area is shown below.

West of Spokane, Washington, between the valleys of the Columbia and the Snake Rivers, the lava plateau has been scoured clean of overlying soils and dissected by deep, sharp-rimmed, braided channels called coulees. One of the largest of these dry channels is now partly filled with the impounded waters of Grand Coulee Dam. An explanation for the origin of this bizarre moonlike landscape was first offered by geologist J. Harlan Bretz in the 1920s. Other geologists initially rejected Bretz's theory as preposterous, but the story he told of the formations of the Scablands is almost universally accepted today. It is a story of ice and water.

The basin of the Clark Fork River in western Montana is rimmed on all sides by mountains. It is drained through a narrow channel around the northern end of the Bitterroot Mountains into the valley of the Columbia River. Sometime during the most recent ice age (ten to twenty thousand years ago) a lobe of the advancing ice [14], nourished by ice fields in the Canadian Rockies, blocked the course of the Clark Fork River where it crosses the narrow northern panhandle of Idaho (see upper illustration at right). This ice dam caused the waters of the Clark Fork and its tributaries to back up, forming a large body of water known to geologists as Lake Missoula.

But ice is a very unstable material for a dam. At some point in time the barrier suddenly gave way and the waters of the lake poured out in a mighty flood that swept across the basalt plateau, stripping and incising the land (see bottom illustration at right). The floodwaters drained at last through the gorge of the Columbia River and thence to the sea. During peak discharge, the volume of water in the brief but terrible surge has been calculated to have equalled that of a hundred Mississippi Rivers. Within a week or two Lake Missoula had drained dry.

This may have been the greatest freshwater flood the surface of our planet has ever seen [39], and it may have happened more than once as the glacier pushed its advancing nose again and again into the channel of the Clark Fork River. This astonishing deluge (or sequence of deluges) left in its wake one of the most tortured landscapes on earth.

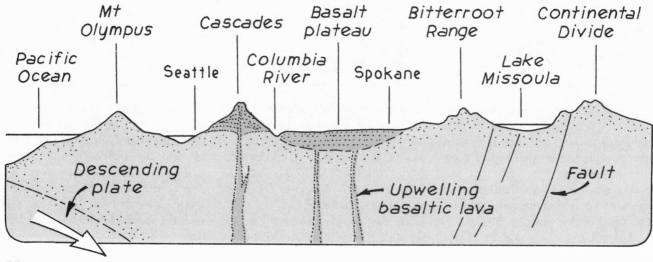

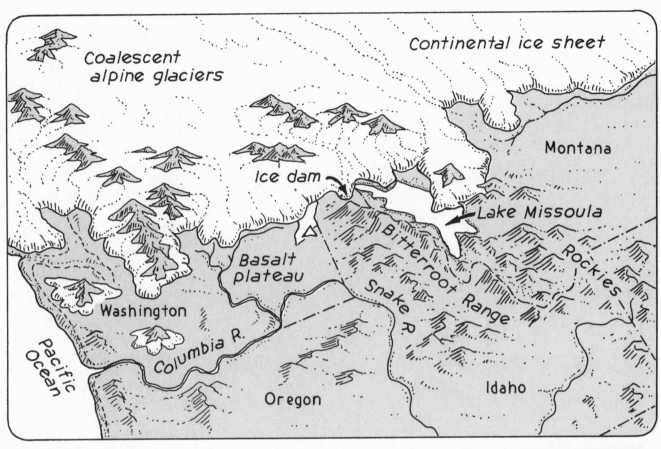

Coalescent alpine glaciers

Continental ice sheet

Ice dam

Montana

Lake Missoula

Bitterroot Range

Rockies

Basalt plateau

Snake R.

Washington

Columbia R.

Pacific Ocean

Oregon

Idaho

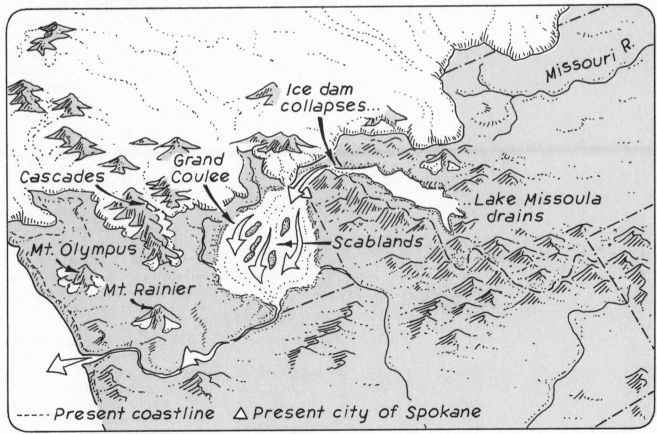

Ice dam collapses...

Missouri R.

Grand Coulee

Cascades

Lake Missoula drains

Mt. Olympus

Scablands

Mt. Rainier

----- Present coastline △ Present city of Spokane

9 Yellowstone Hot Spot

Old Faithful

Indians who lived near the Yellowstone region believed that the area was possessed by spirits, spirits who were adverse that humans should come near. And no wonder! Early trappers who visited the area in search of pelts were astonished by the bubbling multicolored pools and the geysers that shot plumes of steam hundreds of feet into the air. Many of these early explorers described the place with the adjectives of hell, although others found the heat emanating from the ground an unexpected blessing in the cold of winter. It was the remarkable thermal features of the area that led the Congress of the United States to establish the Yellowstone plateau as our first national park in 1872. During the hundred plus years that have passed since the establishment of the park, over 50 million persons have come to gawk at the wonderful manifestations of the earth's internal heat.

In addition to the continuous thermal activity, the Yellowstone region is frequently shaken by earthquakes. Most of the quakes are small, but in 1959 the area just west of the park received a major shock that tore apart the landscape and triggered a massive landslide. The slide buried three dozen campers and dammed the Madison River to form Earthquake Lake.

Most volcanic and seismic activity on the earth's crust takes place along boundaries between the rigid lithospheric plates. Yellowstone, by contrast, is more than 1000 miles from the nearest plate boundary. It is one of several mid-plate centers of geologic activity. The Hawaiian Islands are another [55]. Geologists call these centers of thermal upwelling "hot spots."

A close study of earthquake waves and gravity variations in the Yellowstone region has revealed the existence of a huge reservoir of hot, partly molten rock beneath the park. This colossal column of magma extends from its base a hundred miles deep in the upper mantle to within a few miles of the surface (see illustration on page 30). The column is apparently anchored in the mantle and the North American plate moves over it. The volcanic rocks of the Snake River Plain, which extend like a solid river of rock for 400 miles across southern Idaho, were probably exuded over a period of 15 million years as the continental plate slid southwestward over the hot spot. The earthquake-shaken volcanic plateau of the present Yellowstone National Park is the most recent rattling cover for this geological "teapot."

The geysers of Yellowstone, including Old Faithful (which sprays steam into the air with clockwork regularity), are a product of the thinly covered subterranean chamber of hot rock. Rain water seeps into the ground and is heated by proximity with the magma. It is then forced upward through faults and fissures in the stretched crust to erupt into the atmosphere as hot water and steam, to the recurring delight of tourists.

Yellowstone is quiet today compared to the not so distant past. During the past 2 million years there have been at least three periods of intense volcanic activity. The most recent episode climaxed about 600,000 years ago when two mighty adjacent eruptions expelled more than 200 cubic miles of magma from the subterranean chamber and scattered ash across much of the American west. The roof of the partially drained magma chamber collapsed inward to form a basin-shaped de-pression called a caldera. Almost immediately the collapsed magma chamber began to refill, pushing up the caldera floor and creating two domes over the sites of the twin eruptions.

Eventually, if the North American plate continues to drift westward, the thermal activity now centered at Yellowstone Park will move eastward across Wyoming and Montana, lengthening the thick river of volcanic rock. It is, of course, the continental crust that moves, not the hot spot. In the meantime the surface of the earth over the Yellowstone hot spot continues to rise. The central region of the park has been pushed up several feet during the century that has passed since the park was created. This bulge might be the prelude to another eruption.

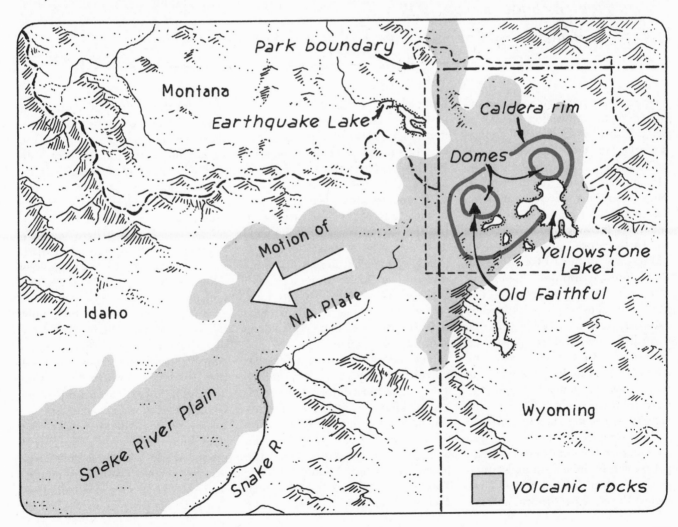

10 The Making of the Rockies

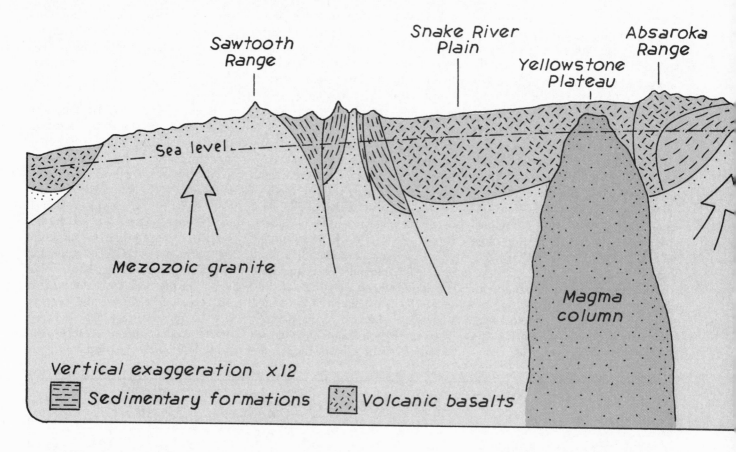

Sawtooth Range

Snake River Plain

Yellowstone Plateau

Absaroka Range

Sea level

Mezozoic granite

Magma column

Vertical exaggeration x12

Sedimentary formations   Volcanic basalts

Geologists are not at all certain of the source of the upwelling column of magma that feeds the Yellowstone hot spot, or how that column remains anchored in the mantle as the lithospheric plate moves over it. These questions still await answers. The geologist who seeks to understand the tectonic forces that shape the earth's crust is something like a physician who must diagnose a patient's internal disorder without benefit of exploratory surgery. The physician relies on clues afforded by external symptoms of the internal disorder—temperature, pulse rate and so on. In the same way, the geologist must deduce the nature of what goes on beneath the earth's crust from clues—heat flow, seismic activity—to be found on the surface.

Few areas of the earth's crust are so replete with tantalizing clues, and yet remain so difficult to understand as the western United States. Geological structures in the arid west are

more clearly exposed than in the east. The west has long been a geologist's playground, and has been closely studied since the Louisiana Purchase in 1803. The establishment of the U.S. Geological Survey in 1879 brought some degree of order to the study of the American west, but the region did not readily yield its secrets. The Yellowstone hot spot is only one geological mystery of the region. The Rocky Mountains offer another. Why there should be a young and growing mountain range so far from a plate boundary is a deep and perplexing puzzle.

The theory of plate tectonics has provided geologists with a unified and comprehensive theory of mountain building. According to the theory, mountain ranges are pushed up by horizontal plate movements, and will generally be found near the boundaries where plates interact. The Andes Mountains of South America [57], the Alps of Europe

[33] and the Himalayas of southern Asia [47] are examples of young mountain ranges whose origins are readily understood in terms of plate motions. All of these ranges lie on present boundaries between converging plates. The squeeze of plates crumples the crust and throws up mountains. Nor is it hard to relate old ranges, such as the Appalachians or the Urals, to plate boundaries where continents collided in the distant past.

But the Rocky Mountains do not lie along a plate boundary. They march down the interior of a presumably stable continent, and yet they evidence powerful thrusting and lifting forces beneath the crust. Geologists generally agree that the lifting of the Rockies, like the stretching and downfaulting that produced the basin and range topography further west, is somehow related to the complex clash of plates that is taking place along the Pacific

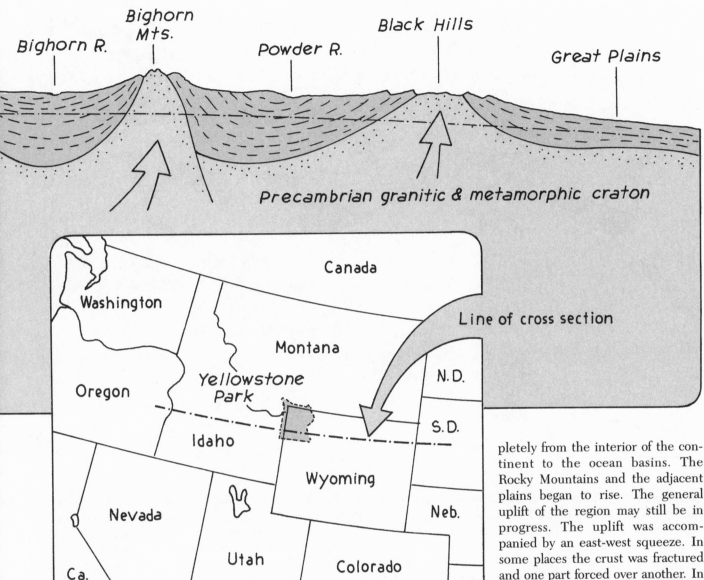

Bighorn R.

Bighorn Mts.

Powder R.

Black Hills

Great Plains

Precambrian granitic & metamorphic craton

Canada

Washington

Line of cross section

Montana

Oregon

Yellowstone Park

N.D.

S.D.

Idaho

Wyoming

Nevada

Neb.

Ca.

Utah

Colorado

coast of North America. But there is no agreement about how the influence of those scraping and colliding plates can be felt so far inland.

Throughout most of the past half billion years of earth history, that part of the continent where the Rockies stand today was a platform of low relief. The region was frequently submerged beneath shallow seas [12]. Thick layers of sediments were deposited on the floors of those seas, providing a blanket of sandstones, limestones and shales over the ancient granite and metamorphic rocks

of the continental crust (the thick block of granitic continental crust is called the *craton*). Then, beginning about 70 million years ago, the area was subjected to the series of thrusts and uplifts known as the Laramide orogeny or mountain-building episode. The revolution may have begun with broad islands folded up in the shallow inland seas of the late Cretaceous period. The basins between these folds received new sediments stripped by erosion from the crests of the uplifts. At last the waters of the shallow seas retreated com-

pletely from the interior of the continent to the ocean basins. The Rocky Mountains and the adjacent plains began to rise. The general uplift of the region may still be in progress. The uplift was accompanied by an east-west squeeze. In some places the crust was fractured and one part forced over another. In other places, such as the Bighorn Mountains of Wyoming, folding and erosion pushed the ancient basement rocks of the granitic craton up through the sedimentary blanket. The Rocky Mountain region is today a place of towering peaks of ancient granite and deep sediment-filled intervening basins. A simplified east-west cross-section is sketched above.

Geologists can only guess at the nature of the forces beneath the crust that have so wracked and warped the American west. Perhaps the clues to a solution are already available, awaiting a new Hutton or Wegener who will unravel the mystery.

11 A Blister on the Crust

Mount Rushmore

The Black Hills of South Dakota rise from the Great Plains of the American west with unexpected abruptness. The summits of the hills have been lifted two to four thousand feet above the surrounding plains, high enough to be covered with dark forest vegetation. It is the contrast of forest to grasslands that gave the hills their name.

The Black Hills stand in stark contrast to the surrounding plains in other significant ways. They are composed of granite and schists, ancient igneous and metamorphic rocks typical of the deeply rooted body of the continent. Elsewhere across the central United States, from the Mississippi to the Rockies, these rocks of the continental craton have been covered with a veneer of sedimentary formations: sandstones, limestones and shales. Many of these sedimentary strata were laid down at a time when the interior of the continent was submerged by shallow inland seas [12].

It was into the finely grained granite face of Mt. Rushmore that Gutzon Borglum, and later his son Lincoln Borglum, between 1927 and 1941 carved the familiar heads of Washington, Jefferson, Lincoln and Theodore Roosevelt. "Carved" is perhaps not the right word. The monument is actually a major work of engineering, involving the removal of 450,000 tons of rock with jackhammers and dynamite. Deep cracks and fissures discovered in the rock as the work progressed required nine separate changes in the original design. The Rushmore monument is wonderfully appropriate as a symbol for the nation's birth and early history. It is located in the heartland of the country, and is shaped from the very stuff and substance of the continent itself, rocks that appear to be at least 2 billion years old.

The heart of the continent of North America, like the other continents, consists of a massive slab of igneous and metamorphic rocks of a granitic composition. This slab, called the craton, is typically 20 to 30 miles thick (thickest beneath the mountains). Like ice on water, the craton floats on the denser and less rigid rocks of the earth's upper mantle. During most of the past six hundred million years, the central part of this flat slab lay near or below sea level. During those times it acquired a blanket of sedimentary formations a mile and a half thick.

Sometime between 70 and 40 million years ago the area of the present Black Hills was locally uplifted, as if

32

a finger were pushing up on the earth's crust from below. The lifting was gentle; there is no evidence of serious faulting or volcanism. As the land rose, the agents of erosion went to work stripping away the sedimentary cover, until at last were exposed the ancient Pre-cambrian granites of the craton. It was into a great outcropping of this granite that the Borglums blasted and chiseled their gargantuan sculpture. The highest of the Black Hills is Harney Peak (7242 feet), another mass of granite that has proved resistant to erosion.

Where erosion encountered particularly resistant strata of the sedimentary cover steep inward-facing ridges were formed, sometimes called "hogbacks." These ridges lie in concentric rings around the Black Hills, with the youngest, uppermost strata composing the outer rings. The oval lowlands between the inner and outer hogbacks were called "the Racetrack" by the Indians. Drainage in the area is governed by geology. Streams tend to flow outward from the uplifted dome, sometimes following the outcrop patterns of the sedimentary rocks, but eventually slicing across the hogbacks to make their way across the plains to the valley of the Mississippi. These features are shown below, in profile and in a bird's-eye view.

The scale of the uplifts that are required to create mountains such as Rushmore are deceptively exaggerated by a drawing such as the profile of the Rocky Mountains on the preceding pages, and in many of the other drawings of this book. Without vertical exaggeration it would be difficult to show topographical details of the earth's crust. The profile of the Black Hills on this page is drawn without vertical exaggeration. It should be clear that the one or two mile high bulge that created the Black Hills represents only a delicate local swelling of the earth's skin. The puzzling origin of the "finger" that exerted the pressure from below to form this thin blister on the crust is part of the more general geological mystery that is the American west.

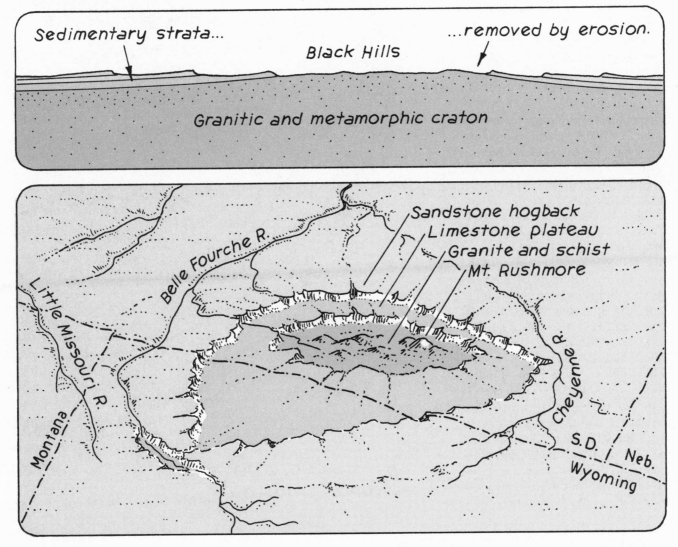

Sedimentary strata... Black Hills ...removed by erosion.

Granitic and metamorphic craton

Belle Fourche R.

Little Missouri R.

Sandstone hogback
Limestone plateau
Granite and schist
Mt. Rushmore

Cheyenne R.

Montana

S.D. / Neb.
Wyoming

12 The Inland Seas

Mt. Rushmore provides a rare glimpse of the granitic foundation of the continent. Except for the region around Hudson Bay in Canada [20], the central part of the North American continent is covered with a blanket of relatively undeformed sedimentary rocks. This blanket is a mile or more thick, and many of the strata it contains are of marine origin. The marine sediments are interleaved with sandstones and mudstones representing debris washed down onto the central platform from now vanished highlands that once stood to the west. These sedimentary strata tell a story of successive inundations of the continental interior by shallow seas.

A rise in sea level seems to have accompanied the splitting up of the supercontinent of Pangaea 200 million years ago. Geologists are not certain how the two events are causally related. It is widely believed that the creation of great mid-oceanic ridges during a period of rapid sea floor spreading diminished the volume of the ocean basins, causing the waters to overflood the continents. No great rise in sea level was required for the oceans to intrude deeply upon the land. Indeed, large areas of the present continents are flooded by the sea. Hudson Bay in Canada is an example of a shallow continental sea. The proportion of dry land to sea on the surface of the earth is very delicately balanced and subject to dramatic change. Rates of sea floor spreading or continental crumpling can cause changes in sea level by affecting the volume of ocean basins. Changes in climate, which affect the amount of water stores in continental ice caps, such as those on Greenland and Anarctica, can also cause substantial changes in sea level. Climatic change may itself be related to the changing arrangements of continents (among many other factors), but geologists have a long way to go before they fully

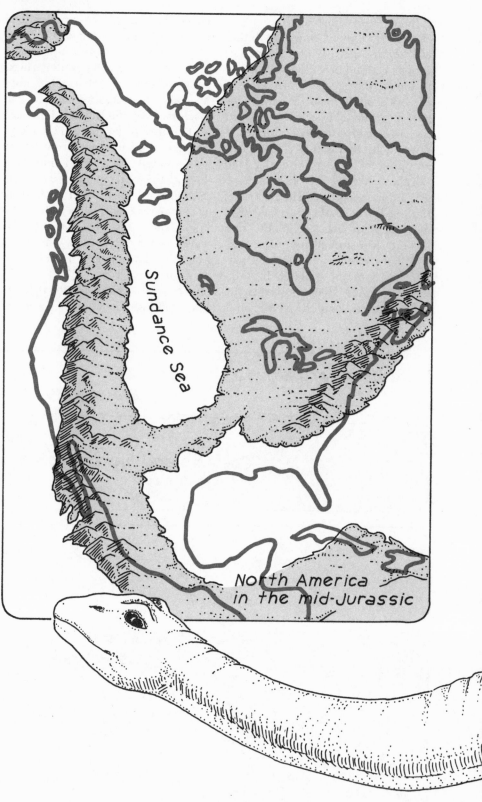

Sundance Sea

North America in the mid-Jurassic

Apatosaurus, a Jurassic sauropod dinosaur

understand how these great tectonic and climatic systems are related.

By mid-Jurassic times (150 million years ago) the oceans had intruded far onto the interior of the North American craton. A great tongue of the Arctic Ocean known as the Sundance Sea covered much of the region where the Great Plains and the Rocky Mountains are today (see map). The climate was milder then than now, perhaps because the extended sea surface had a moderating influence on climatic change. The swampy margin of the great inland sea was an ideal habitat for the giant reptiles, the dinosaurs, that dominated the earth during the Jurassic era. Illustrated on these pages is Apatosaurus, one of the greatest of the "thunder lizards," similar in most respects to the more familiar Brontosaurus. Even with their massive pillarlike legs, these great beasts must have welcomed the buoyant waters of the inland seas to help support their weight. Toward the end of the Jurassic, the Sundance Sea retreated northward, leaving behind a broad swampy plain. The strata of mud, gravel and silt laid down during this period have proven to be a rich lode for fossil hunters in search of the bones of dinosaurs. The skeletons of the sauropods, the largest of the dinosaurs, are common in rocks of the late-Jurassic period, which were laid down on the bed of the retreating Sundance Sea.

A new inundation of the continents occurred about 100 million years ago. This time the waters advancing from the north met the waters overflooding the continent from the south. All of the central and southeastern United States was covered in one vast expanse of water. Eastern Canada and the Appalachians stood as island bastions on the submerged continent. In the west the sea lapped the feet of highlands which were the precursors of the Rockies. Rich beds of coal were laid down along the swampy margin of this extensive inland sea. Those beds of coal today constitute one of America's most valuable mineral resources.

But the days of the shallow seas, and of the dinosaurs [34], were drawing to a close. The western margin of the continent was being slowly crumpled eastward by the push of the Pacific plate, in a way not yet fully understood, and at last those forces began to thrust up the present Rockies from the floor of the old Sundance Sea. The waters of the great inland flood retreated before the slow "avalanche" of sediments washed down into the seas from the new uplands. At the same time, the fossils of swamp-loving dinosaurs were lifted to the peaks of the new Rocky Mountains.

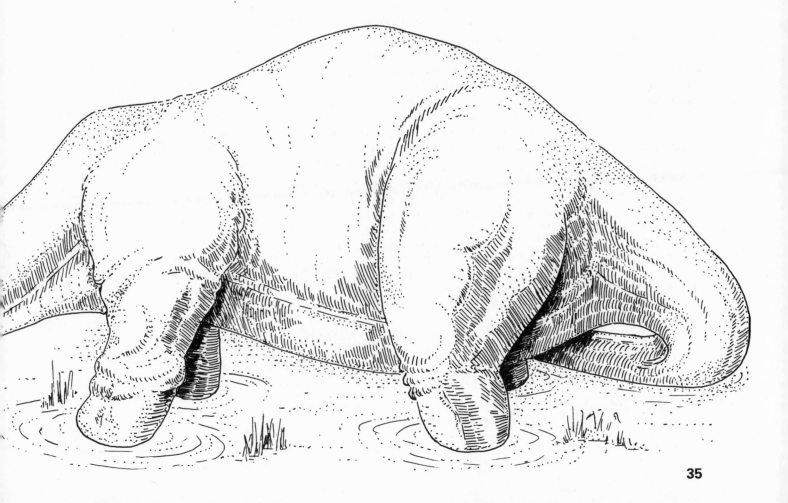

13 The Greatest Quake

"The ground rose and fell as earth waves, like the long, low swell of the sea, passed across its surface, tilting the trees until their branches interlocked and opening the soil in deep cracks as the surface was bent... On the Mississippi great waves were created, which overwhelmed many boats and washed others high upon the shore, the return current breaking off thousands of trees and carrying them out into the river, sandbars and points of islands gave way, and whole islands disappeared..."

This is how the quake was described by a seismologist reconstructing the events of 1811-12. The quake did not occur in the tortured and broken American west, where the vast majority of U.S. earthquakes are felt, but in the stable central lowlands. It was the greatest earthquake to shake the United States in historic times, and possibly one of the greatest ever to rock the crust of the earth.

The epicenter of the quake was near New Madrid, Missouri, a sleepy frontier town on the banks of the Mississippi. Missouri had only recently become part of the United States; the quake followed the Louisiana Purchase by only eight years. The area most severely affected by the earthquake was sparsely populated and loss of life and property was slight. If the same earthquake were to occur today, it would be a tragedy of staggering proportions. Cities of the central Mississippi valley, such as St. Louis and Memphis, would suffer extensive damage.

The New Madrid earthquake was rated XII on the modified Mercalli scale of intensity. XII is the highest rating on that scale. The scale describes a level XII quake in this way: "Super panic. Damage total. Waves seen on ground surfaces. Lines of sight and level distorted. Objects thrown upward into the air." By contrast, the great San Francisco earthquake of 1906 was rated XI.

The New Madrid event actually consisted of three separate shocks of the maximum intensity. The first shock occurred during the night of December 16, 1811, and continued with diminishing intensity throughout the next few days. On January 23, 1812 there was a further shock, as strong as the first, followed by two weeks of quiet. A last furious jolt was felt on February 7. The course of the Mississippi River was changed and two large lakes, St. Francis and Reelfoot, were created in basins of down-dropped crust. Trunks of cypress trees drowned a century and a half ago still stand like grey ghosts in the black waters of Reelfoot Lake. Minor aftershocks of the quakes lasted for two years. It is said that vibrations from the quakes woke sleepers in Washington, D.C. and rang church bells in Boston. As the map at right shows, the intensity of the quake was VII even as far away as Chicago and Detroit, sufficient to cause alarm among the inhabitants.

Most earthquakes in the earth's crust are associated with the divergence or convergence of the lithospheric plates. Global maps showing the epicenters of earthquakes over an extended period of time almost perfectly outline the plate boundaries. The New Madrid earthquake is an example of the rarer mid-plate quake. It reminds us that plate rifting and collision have been going on throughout the long history of the earth. Deep beneath the Mississippi valley there are old faults and seams associated with earlier episodes in the breaking apart and welding together of continents. In 1811-12 near New Madrid, Missouri, one of those ancient faults readjusted itself like an old house that creaks as it settles, possibly to compensate for the loading of the crust with sediments carried down from the western highlands by the Missouri and Mississippi Rivers. Charleston and Boston are other eastern localities that have felt severe earthquakes in historic times. Not even the geologically stable eastern seaboard is immune to the possibility of a shock.

Every year hundreds of thousands of earthquakes are recorded on the crust of the earth. Most are too feeble to be felt by humans. Only a few might approach the intensity of the New Madrid quake. Earthquakes annually kill an average of more than 10,000 people. While some places on the earth's crust are particularly at risk, the crust is a graveyard of old faults and no place is truly safe.

Boston 1755												
Charleston, 1886 - Yellowstone, 1959 - Anchorage, 1964												
San Francisco, 1906												
New Madrid, 1811-12												
I	II	III	IV	V	VI	VII	VIII	IX	X	XI	XII	

Earthquake Intensity Scale

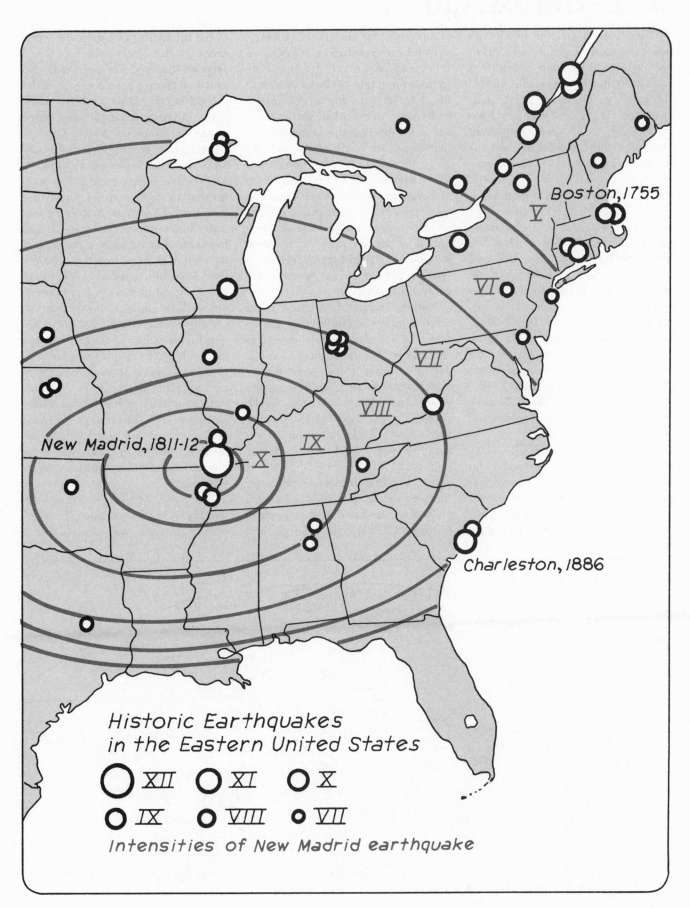

New Madrid, 1811-12

Boston, 1755

Charleston, 1886

V

VI

VII

VIII

IX

X

Historic Earthquakes
in the Eastern United States

⬤ XII ◯ XI ◯ X

◯ IX ◯ VIII ◦ VII

Intensities of New Madrid earthquake

14 The Ice Age

By the beginning of the nineteenth century, mountaineers who lived hard against the glaciers of the Swiss Alps had come to the conclusion that those fingers of ice had in the past extended far greater distances down their valleys. The first important champion of this radical idea was the naturalist Louis Agassiz. Deep parallel scratches on widely separated outcrops of bedrock and large boulders stranded on high ground far from their apparent bedrock sources convinced Agassiz that much of the region of the Swiss Alps had once been covered by an extensive sheet of ice, of which the present valley glaciers were only a paltry remnant. That ice had been part, he declared, of a great polar ice cap that blanketed much of Europe and North America during a prehistoric "Ice Age." When Agassiz presented his ideas to the scientific community in 1837, they were met with almost universal ridicule.

By the second half of the nineteenth century, explorations in the north and south polar regions had made geologists aware of the extent of the Greenland and Antarctic ice sheets, and Agassiz's extravagant

claims of ice sheets in Europe seemed somewhat more reasonable. But more important, by that time the accumulation of direct local evidence for ice caps on parts of Europe and North America had become persuasive. Once people knew what to look for, the evidence of massive glaciation was to be seen all around. The shapes of valleys, the disposition of boulders, scratches on outcrops, and depositional patterns of erosion materials all confirmed the startling notion of an ice age. As so often happens in the history of science, a "preposterous" idea became accepted doctrine.

Various reasons have been given for the onset of continental glaciation, and we shall look at several of these later. The most recent ice age (we now know there were earlier ones) began about 125,000 years ago when the climate of the earth cooled. At the onset of the cooling the extent of glaciation in North America may have been no greater than today (see drawing 1 below). In addition to the Greenland ice sheet, there may have been smaller ice caps on Baffin Island and other islands of northern Canada, and alpine glaciers

in the mountains of Alaska and western Canada. The Great Lakes and Hudson Bay did not yet exist, and much of the region now drained by the Missouri, Ohio and St. Lawrence Rivers drained northward across the Canadian plains.

Only one condition is necessary for a glacier to form anywhere on earth—more snow must fall in winter than melts in summer. Slowly the thickness of the snow mass builds up until finally it is turned into ice and begins to flow under the pressure of its own weight. Mountain glaciers flow down hill, assisted by the tug of gravity. Continental glaciers squeeze out at their margins, slowly increasing their area. By 100,000 years before the present, substantial ice caps had built up in Canada on either side of the lowlands that would eventually become Hudson Bay (2). In the west, mountain glaciers had begun to coalesce to fill the valleys between peaks. As the ice expanded its dominion, old river channels were blocked and drainage patterns began to change.

The ice reached its maximum extent about 18,000 years ago (3). All of eastern Canada was covered by a

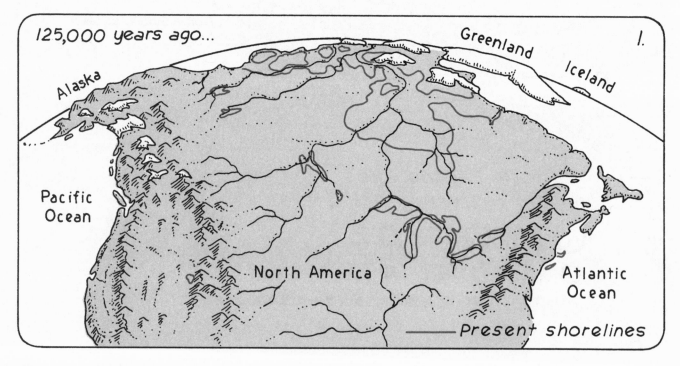

125,000 years ago...   Greenland   1.
                                                                                Iceland
Alaska

Pacific
Ocean

                                                                                Atlantic
                                                                                Ocean
                    North America
                                                        ——— Present shorelines

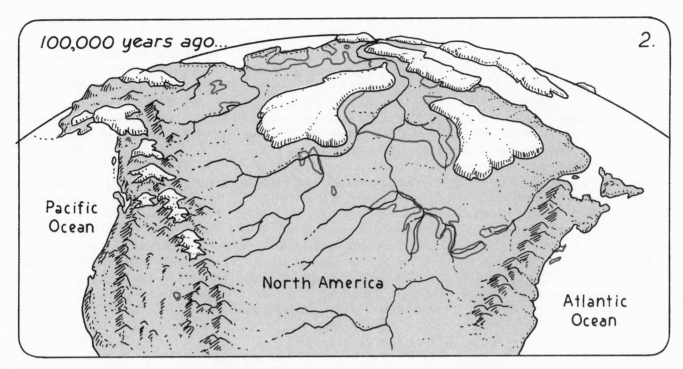

100,000 years ago...

Pacific
Ocean

North America

Atlantic
Ocean

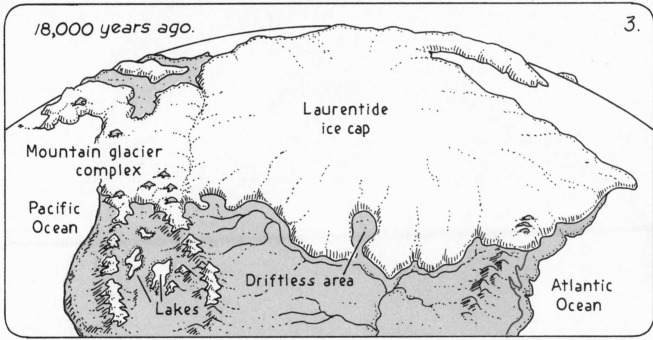

18,000 years ago.

Laurentide
ice cap

Mountain glacier
complex

Pacific
Ocean

Driftless area

Lakes

Atlantic
Ocean

single frozen mantle, called the Laurentide ice cap. Glaciers in the western mountains had fused together and merged with the Laurentide sheet to form a continuous expanse of ice that reached from coast to coast.

The ice advanced in great lobes, probing along the lowlands, pushed forward by the weight of the ice behind. Where the spreading glacier met the sea it tried to push out onto the water, but on this unstable base it broke apart into blocks that floated away as icebergs. On land the advancing glacier halted only when it had pushed so far south that melting at its southern terminus kept pace with the inflow of ice from the north. The ice of the Laurentide glaciation came as far south as the present courses of the Ohio and Missouri Rivers. Indeed, those river systems were created when the old northern drainage channels were closed by the ice. The channels of the new rivers were cut and deepened by the great flow of meltwater that the rivers carried off to the south from the melting margins of the ice.

15 Creatures of the Ice

The great ice cap that blanketed northern lands between ten and seventy-five thousand years ago was only the most recent of a succession of ice ages which have occurred during the past few million years. Indeed, throughout these several millions of years the earth's climate has been generally cooler than today, and the entire long period is sometimes referred to as the "Great Ice Age." The Great Ice Age coincided with the rise of humans to prominence on the planet; in a real sense, *homo sapiens* is a creature of the ice [31]. We are apparently still in the Great Ice Age, although in something of an intermission, and large portions of the earth's crust (Greenland, Antarctica) remain burdened with ice. The Great Ice Age is not unique. Other long regimes of subnormal temperatures have affected the planet throughout its history, recurring at intervals of hundreds of millions of years.

What is the cause of these great chapters of continental glaciation? There has been no scarcity of theories to explain the required variations in the earth's climate, but there is little general agreement as to which theory is correct. The theories fall into several categories: 1) Variations in the output of radiation by the sun. Perhaps the sun is not as constant a star as we have believed in the past. 2) Changes in the earth's atmosphere that keep the sun's radiation out or the earth's heat in. For example, variations in the levels of carbon dioxide, ozone or volcanic ash in the atmosphere would have one or the other of these effects. 3) Variations in the distribution and intensity of the sun's radiation over the surface of the earth as a result of periodic astronomical changes in the position of the earth relative to the sun. This class of theories has recently received much attention and has to its credit rough correlations between predicted and actual climatic variations. 4) Lateral movements of the continents with respect to each other or to the poles, as described by the theory of plate tectonics. Plate motions could affect the general elevation of the continents with respect to sea level, or change

Woolly
Mammoth

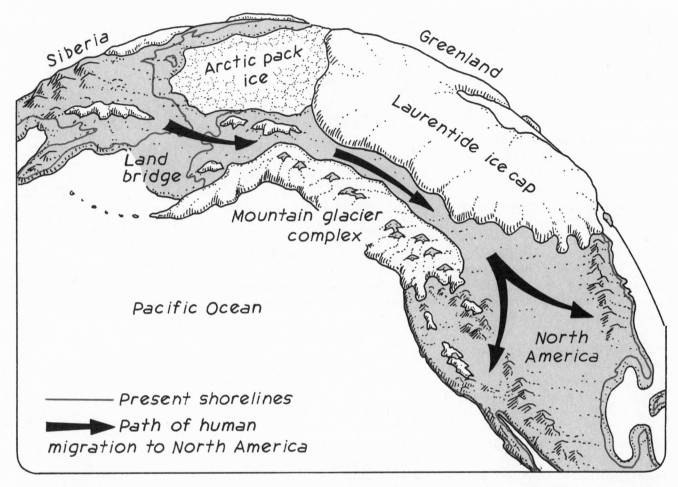

Siberia

Arctic pack ice

Greenland

Laurentide ice cap

Land bridge

Mountain glacier complex

Pacific Ocean

North America

——— Present shorelines

➤ Path of human migration to North America

circulation patterns in the oceans and atmosphere.

Theories in the latter class are relatively new but are coming on strong. Geologists have become acutely aware of the dynamic nature of the earth's crust. The earth's surface systems of land, water, atmosphere and ice are components of a complex and delicately balanced machine, all closely linked to crustal change. A time-lapse movie of the earth's surface made from the moon over the past four billion years would show continents colliding and rifting apart, oceans' basins being consumed and created, mountains being thrust up and worn down. It would be surprising if these dramatic tectonic events did not result in significant climatic change.

The most recent glaciation of the Great Ice Age, the glaciation described on the preceding pages, is of

course the one we know the most about. Many of the geographical features we see in the northern and western United States were shaped by that last advance of the ice (7, 8, 16). The effect of the ice on living things was also dramatic. Consider the animal depicted at the left, the wooly mammoth. This twelve-foot high relative of the elephant was common in North America throughout the last ice age. The creature vanished during the glacier's stately retreat, only 10,000 years ago. The ice age mammoths that ranged the Americas had the company of mastodons, saber-toothed cats, huge camels, wild horses and other beasts. All of these creatures vanished together. The cause of the mass extinctions during the closing stages of the ice age is not known. Perhaps the wooly mammoth and its exotic companions were

simply unable to adapt to the increasingly temperate environment that accompanied the retreat of the ice.

In North America the mammoth was certainly helped toward extinction by the arrival of human hunters. Humans seem to have arrived in North America sometime between twelve and thirty thousand years ago. The drop in sea level which accompanied continental glaciation created a land bridge across the Bering Strait between Asia and Alaska. During several periods of temporary warming, a narrow corridor opened between the Laurentide ice cap and the western alpine glacier complex. Across this bridge and along this corridor the first Americans migrated from Siberia. The last wooly mammoth may have fallen victim to the sharp flint-pointed spears deftly thrown by the new inhabitants.

41

16 The Making of the Great Lakes

The ice sheet that blanketed much of North America only 10,000 years ago was in the areas of maximum accumulation more than a mile thick. During the tens of thousands of years this mighty cloak of ice lay upon the shoulders of the continent it was moving ice, a massive grinding bulldozer of ice moving across the landscape. Everywhere the glacier lay, its work is evident today. Valleys were scooped out and rounded by the moving ice; peaks were scraped clean. Huge quantities of rock were torn from the northern lands and carried south. Long high east-west ridges of this eroded debris were deposited by the ice at its melting southern margin [17]. Further, the weight of the huge mass of ice depressed the crust of the earth, in some parts of Canada by over a thousand feet. The crust is still rebounding from that great depression.

Perhaps the most conspicuous features of the post-glacial landscape are the Great Lakes on the border between the United States and Canada. The lakes are the greatest freshwater resource in North America, with a combined surface area greater than any other freshwater lake on earth. No other large freshwater body lies at such favorable latitudes. The history of the making of these lakes is long and complex, but the key episodes are told by the drawings on these two pages.

As the continental ice sheet pushed down from its primary centers of accumulation in Canada, it moved forward in lobes of ice that followed the existing lowlands (1). Before the coming of the ice, the basins of the present Great Lakes were simply the lowest lying regions of a gently undulating plain. These low lying areas corresponded to the places where easily eroded strata of sedimentary rock were exposed at the surface. The moving tongues of ice scoured and deepened these lowlands as the glacier made its way toward its eventual terminus near

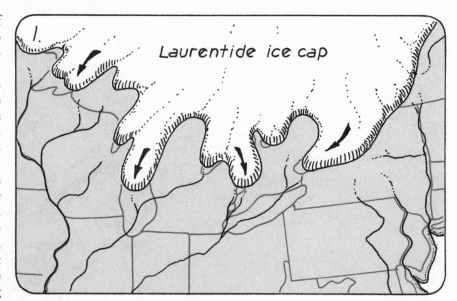

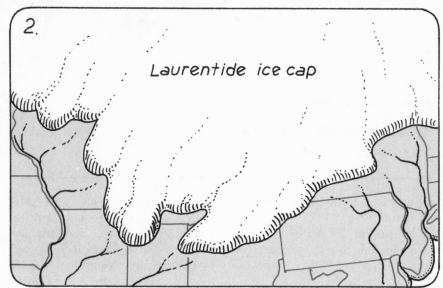

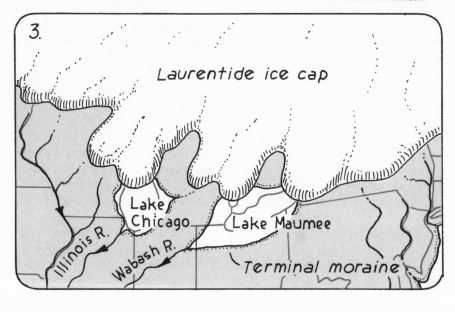

the present Ohio and Missouri Rivers.

About 16,000 years ago the ice sheet stood for a long time with its margin just to the south of the present Great Lakes (2). Erosional debris carried by the moving ice was dumped at the melting southern edge of the glacier and built up long ridges called terminal moraines. When the ice began to melt back from this position about 14,000 years ago, meltwater collected behind the dams formed by the moraines (3). The crust behind the moraines was still depressed from the weight of the ice it had borne, and this too helped create the ancestral Great Lakes. The first of these lakes drained southward across Illinois and Indiana, along the channels of the present Illinois and Wabash Rivers.

As the ice cap continued to retreat, and the exposed crust rebounded from the weight of the ice, the meltwater lakes found new outlets to the sea (4). About 12,000 years ago the eastern lakes found an outlet through the valleys of the Mohawk and Hudson Rivers across New York.

A further stage was reached when the ice at last uncovered the St. Lawrence estuary (5). Because the crust was still depressed, that arm of the sea was wider and deeper than at present, and one branch of the estuary extended up the lowland valley where Lake Champlain is today. At this stage, all of the Great Lakes found an outlet to the Atlantic along the St. Lawrence. In the east, Lake Erie poured across the Niagara escarpment into Lake Ontario, and thence to the St. Lawrence. The western lakes found an outlet across Canada along the depressed channel of the Ottawa River. Because this channel was still greatly depressed by the ice, the western lakes stood at lower elevations and were less extensive than at present.

When the ice completed its slow retreat to the Arctic islands of the north, the Ottawa River channel re-

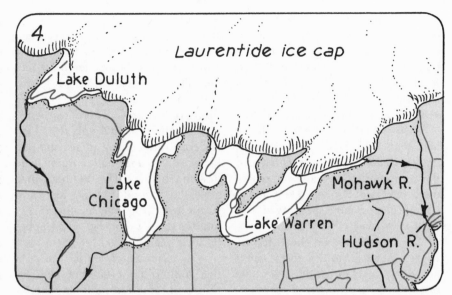

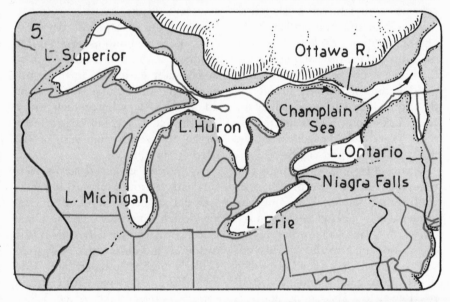

bounded and the Great Lakes assumed their present configuration.

All across central Canada, from Lake Superior to the Arctic Ocean, other extensive glacial lakes were created in the same way. A glance at a map of Canada will reveal their shrunken remnants. The greatest of these ice age lakes is known as Lake Agassiz, a vast expanse of water that once covered much of the Canadian province of Manitoba. The present Lakes Winnipeg and Manitoba are two relics of that inundation.

The story told here of the making of the Great Lakes illustrates one

remarkable fact about planet earth: to use a crude metaphor, the planet is more like a flexible child's balloon than a rigid bowling ball. As rock, water or ice is redistributed on the surface of the earth, the surface rises or sinks to maintain a kind of floating equilibrium, in the same way that a raft sinks deeper in the water as a passenger steps on or rises when weight is removed. This gentle warping of the crust is but a tiny fraction of the earth's diameter, but it can represent dramatic topographical modifications on the human scale.

17 The Making of Cape Cod

Nowhere has the ice left a more delicate and felicitous trace than at Cape Cod, Massachusetts, and the offshore islands of Nantucket and Martha's Vineyard. The "bare and bended arm of Massachusetts," Thoreau called Cape Cod. And that is exactly what the Cape looks like: a bare and unlikely limb of sand and gravel, defiantly shaking its fist and flexing its muscle against the crashing breakers of the sea. The Cape is young; geologically speaking it was born yesterday.

The Laurentide ice sheet, fed by winter snows far to the north, could not keep up forever its relentless push south. Eventually it found its way into a climate warm enough so that melting and evaporation at the glacier's southern lip kept pace with the inflow of ice from the north. On the eastern seaboard of North America that line of furthest advance of the expanding ice sheet fell along New York's Long Island and the offshore islands of Massachusetts (see line 1 on the map at right).

Again, two important points to remember: 1) The North American continental glacier, like mountain glaciers, was *moving* ice, squeezed outward from its centers of accumulation in Canada. 2) This was dirty ice, carrying a load of eroded debris, from giant boulders to sand-sized particles. The ice moved like a bulldozer across the landscape, crushing forests, scouring the land, grinding smooth the summits of hills, scratching at the walls of valleys, plucking boulders from the lee sides of outcrops. All of this huge volume of eroded material was ground and crushed beneath the ice, deposited beneath the ice as glacial "till," or carried by the ice to its southern terminus. There, as the ice melted, the glacier dropped its load of debris into an ever increasing pile known as a terminal moraine (see drawing below). The lighter sands and silts carried by the ice were swept away from the ice margin by meltwater to form a gently sloping outwash plain.

The terminus of the ice stood at this southern limit (line 1 on map) for thousands of years, building its moraine and sandy outwash plain. At that time the ice stood on dry land, and wooly mammoths foraged the tundra south of Nantucket. With so much water locked up in the continental glaciers, the sea level stood hundreds of feet lower than today, and the New England coastline lay far to the south and east of the present coastline.

As the ice advanced across New England, it moved forward in lobes, as it did further west along the valleys which would become the Great Lakes. Where two lobes met, the terminal moraine was built particularly high. Nantucket and Martha's Vineyard were built up at vertices between lobes of ice. When the ice retreated and the sea rose, these high points remained above the waters even as the connecting ridges were submerged.

But before the sea returned, the climate warmed slightly and the glacier retreated to a second standing line (line 2 on the map). Here it again remained more or less stationary for thousands of years. And here again it built up a long ridge of sand, gravel and boulders, material carried in some part all the way from northern New England and Canada. It was this second moraine that became the backbone of Cape Cod and the Elizabeth Islands. Other lobes of ice built moraines on the exposed continental shelf to the east of the present Cape, but those delicate ridges were submerged and erased by the rising sea at the end of the ice age.

By 10,000 years ago the ice was in full retreat and the sea returned as the level of the world's oceans rose. Immediately the wind and water began their energetic work of modifying the "fragile outposts" left behind by the ice. Much of the outer arm of the Cape was carved away by the action of the waves and redeposited in long bars and hooks of sand, such as those near Provincetown.

Today, the Mid-Cape Highway runs along the rocky hump of the old moraine (see profile below right). To the south of the highway, a sandy outwash plain stretches away to meet Nantucket Sound (and, of course, continues beneath it). The summer bathers who frolic on the beautiful beaches where outwash plain meets the sea are enjoying a splendid gift of the ice.

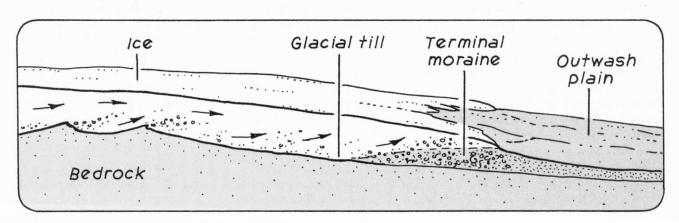

44

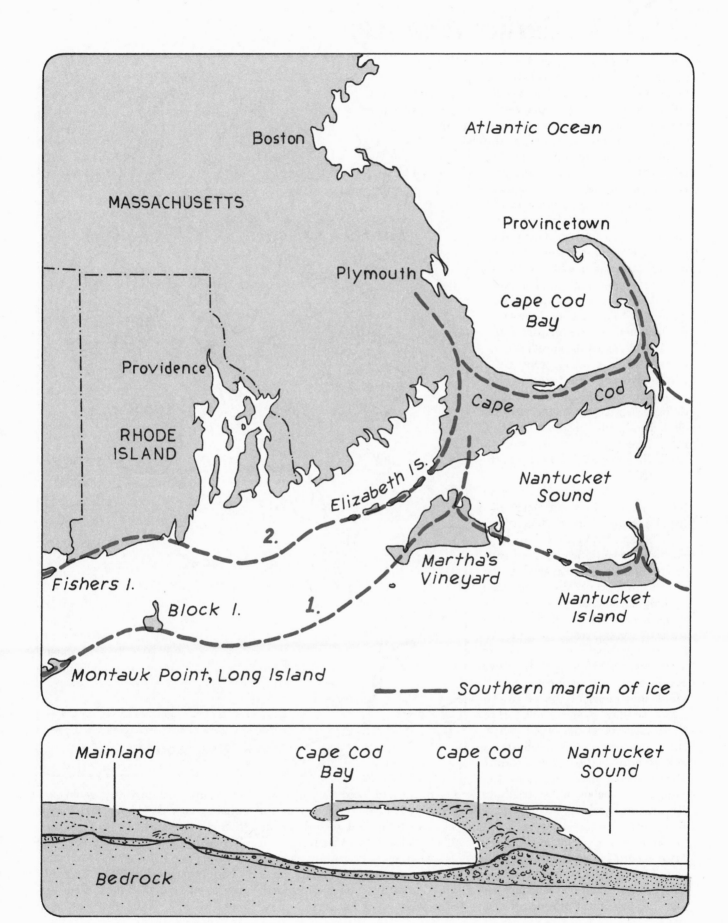

Boston

Atlantic Ocean

MASSACHUSETTS

Provincetown

Plymouth

Cape Cod
Bay

Providence

Cape   Cod

RHODE
ISLAND

Nantucket
Sound

Elizabeth Is.

2.

Martha's
Vineyard

Fishers I.

Nantucket
Island

Block I.

1.

Montauk Point, Long Island

– – – – Southern margin of ice

Mainland

Cape Cod
Bay

Cape Cod

Nantucket
Sound

Bedrock

18 An Earlier Atlantic

The theory of plate tectonics and continental drift has provided solutions to many long-standing puzzles posed by the rocks of the earth's crust. Consider, for example, the distribution of fossils from the Cambrian and early Ordovician periods, 600 to 500 million years ago. Paleontologists have long recognized that life forms from that time can be divided into several fairly distinct and stable provinces or family relationships. One province can be roughly associated with the North American continent, and another with the European and African land masses. A trilobite of each province is illustrated on the map at right. The trilobites, bottom-dwelling creatures of shallow waters, were one of the dominant creatures of the Cambrian period. The two roughly defined faunal provinces also included mutually distinct forms of other invertebrate species, such as corals, brachiopods, and early freshwater fish.

It is not, of course, puzzling to find variations between faunal distributions which are separated by a wide ocean. But wait! The rocks of northern Ireland, Scotland and eastern Scandinavia contain fossils of the "North American" kind, separated by only an imaginary line from the rocks bearing fossils of the "Euro-African" variety. And across the Atlantic the fossil-bearing rocks of southern Newfoundland, Nova Scotia and eastern New England yield specimens of the "Euro-African" families of ancient life. How these enclaves of out-of-place creatures managed to establish themselves across the wide Atlantic Ocean was a great mystery. Certainly, the shallow-water trilobites lacked the capability to traverse a deep ocean basin.

But the story contained in the fossil record has not yet been fully told. During late Ordovician and Silurian times, about 450 million years ago, the two faunal provinces began

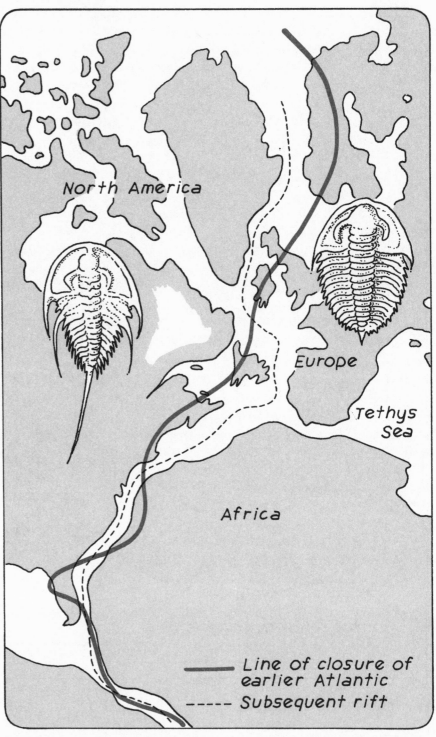

Line of closure of earlier Atlantic
----- Subsequent rift

to converge and loose their distinctness. Fossils in 400 million year old rocks show identical records of life on both sides of the Atlantic.

It would seem that we have here a complicated story of the migration of living creatures, back and forth across wide water barriers. Not so, say some contemporary geologists; what we have is evidence for a migration of continents.

According to the new theory, the dividing line between the two faunal distributions (colored line on maps) represents the shorelines of an earlier "Atlantic" ocean that existed in

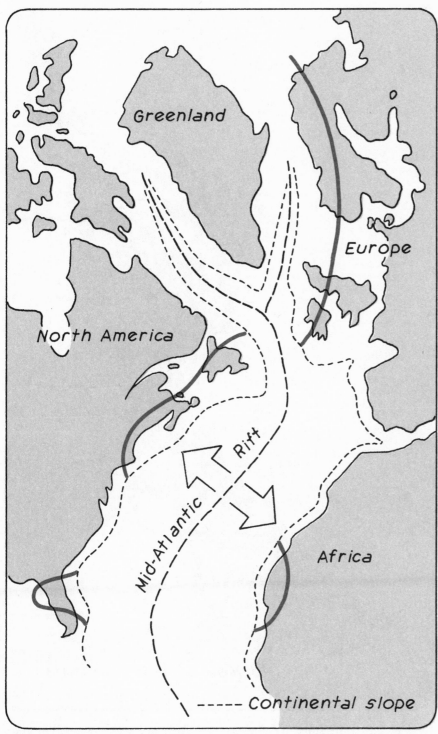

Greenland

Europe

North America

Mid-Atlantic Rift

Africa

- - - - - Continental slope

distinct creatures simply disappeared!

The elimination of the old "proto-Atlantic" ocean floor by subduction beneath the converging continents, and the great crunch that raised the mountains, welded the eastern and western continents together into a single land mass (see map opposite page). But this supercontinent, called Pangaea, did not endure for long. Rifting soon split apart the new continent, and drift—this time in the opposite direction—began to open up the present Atlantic Ocean (see map this page).

But the line of rifting where the new supercontinent broke apart did not exactly correspond to the old seam where colliding continents had been welded together. Scotland, northern Ireland and eastern Scandinavia are parts of the old North American continent that became affixed to Europe. And parts of maritime Canada and eastern New England were "borrowed" from Europe and Africa. Boston, Massachusetts, sits on rocks that were once a part of North Africa, and somewhere between Boston and Springfield, Massachusetts, is the line marking the disappearance of an ocean and the fusion of one continent to another.

In the destruction of one ocean and the creation of another, the continents exchanged fragments along their margins. With the exchanged rocks went the fossil evidence of ancient life forms, and the creation of a puzzle for the fossil hunters. The puzzle awaited the theory of plate tectonics for a solution. The complete story of the disappearance of the proto-Atlantic Ocean and the growth of the present ocean is told on the following pages. Bear in mind, as you read, that the record of the rocks and the fossils they contain is not easily interpreted. Appalachian geology is especially complex. Some geologists will take issue with details of the story told here.

Cambrian times. The shorelines of that ocean did not exactly correspond to those of the present Atlantic. About 500 million years ago, the ancient "proto-Atlantic" began to close up, bringing the continent of North America into contact with Europe and North Africa. It was this collision of continents that gave rise to the Appalachian and Caledonian mountain ranges (see next pages). It also brought about the convergence of the two distinct faunal provinces by bringing their separate habitats into proximity. The wide ocean barrier between the two provinces of

19 The Big Crunch

Today the eastern seaboard of North America is geologically quiet. The slow, patient work of weather and water continues to cut down the highlands. In another few hundred million years the Appalachian states will be as flat as Kansas. The eroded material is transported by rivers to the sea, and great wedges of sediments are built up on the continental shelf and slope. This ongoing business of weathering and transportation of sediments is a quiet activity, the almost imperceptible work of the cycle of the seasons. Only rarely does a minor shudder of the crust disturb the geological peace of the eastern seaboard, as some ancient fault deep beneath the surface readjusts itself like the shift of a sleeper.

But the geologist knows the east has not always slept. First of all, there is the evidence of the Appalachians themselves, old mountains obviously created by a violent forcing and folding of the earth's crust. Then there are the great masses of volcanic rock, such as the thousand foot thick sill of lava that outcrops in the vertical bluff of the Hudson River Palisades (see drawing page 51). And there are the great outcrops of granite, such as those which give the "Granite State" of New Hampshire its nickname. We know with some certainty that granite is formed from intrusions of magma that slowly cool and solidify far beneath the earth's surface, and yet the granite face of New Hampshire's "Old Man of the Mountains" looks benignly down from a lofty peak. Some great uplift of the crust must have placed him there. These things suggest a long history of fire and violence.

John Townsend Trowbridge, in his famous poem "The Old Man of the Mountains," asks: "What earthquake shaped, what glaciers scraped, That nose and gave the chin its angle." Geologists have asked the same question and an answer is beginning to emerge from the theory of plate

Old Man of the Mountains

tectonics. It is the story of the making of the Appalachian Mountains, of which New Hampshire's "Old Man" is a stunning cap. The story is long and complex. But the key episodes of that story are summarized by the drawings on these four pages. Remember that the vertical scale in the drawings has been exaggerated.

The story begins 600 million years ago, when the coast of North America was not unlike the coast of today, except that in most places the actual edge of the continent was further to the west (with respect to the bulk of the continent) than the present margin (1). Separating North America from Africa and Europe was an earlier "Atlantic" ocean, sometimes called the "proto-Atlantic." The continental slopes of this ancient sea and the adjacent ocean basins carried thick loads of sediment which had been washed down from the eroding continents. On opposite continental shelves of the proto-Atlantic lived the distinct provinces of life, trilobites, corals, brachiopods and early fish, described on the preceding pages.

Then, about 500 million years ago, the proto-Atlantic began to shrink, decreasing in width at the rate of an inch or so a year. The basaltic sea floor was pulled or pushed down beneath the eastern margin of North America, in much the same way as the Juan de Fuca plate is being subducted today in the Pacific northwest (2). The possible causes of these horizontal displacements were discussed in the Intro-

duction. Heat generated along the diving plate caused local melting and granitization of crustal rocks. The heating may also have given rise to a string of volcanic islands parallel to the coastline, not unlike the island arcs that now ring the western Pacific. The squeezing pressure of the converging plates and the heat generated by subduction metamorphosed the thick wedge of offshore sedimentary rocks. The squeeze also

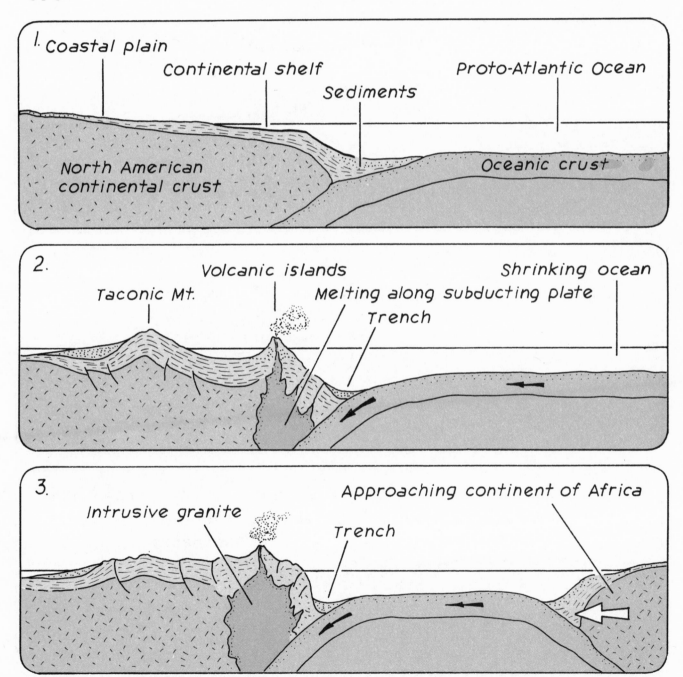

(Continued)

crumpled up the high mountains of which the present Taconic Range along the eastern border of New York is the eroded remnant. The region was wracked by powerful earthquakes as the sea floor plate dove to destruction in the mantle.

The subduction of the ocean floor continued, and the margin of the continent was further compressed (3). Granite and metamorphic rocks were created deep beneath the sur-face, rocks that would become new continental crust. Volcanoes erupted in New England, stoked by the same source of heat that presently fires Mt. St. Helens. The waters of the shrinking proto-Atlantic, caught between converging continents, made their way to other ocean basins. Meanwhile, Africa and Europe rode the subducting plate ever closer to North America.

At last came the big crunch, about 250 million years ago. Africa and Europe slammed broadside into North America—although not quite simultaneously—crumpling the crust and raising a mighty range of mountains (4). These mountains consisted of metamorphosed and folded sediments from the old continental margins and had intruded cores of newly formed granite.

Immediately, erosion began its work of cutting these lofty Alplike

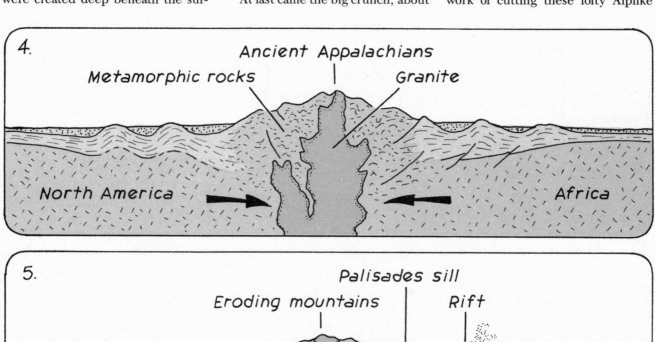

4. Metamorphic rocks — Ancient Appalachians — Granite

North America — Africa

5. Eroding mountains — Palisades sill — Rift

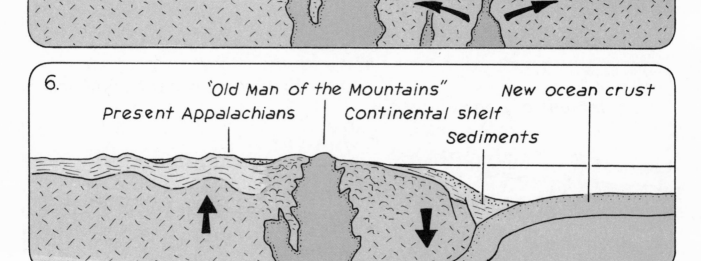

6. "Old Man of the Mountains" — New ocean crust
Present Appalachians — Continental shelf — Sediments

Hudson River Palisades

mountains down, and the debris was deposited in thick wedges to either side of the range. Not long thereafter, about 180 million years ago, compression in this part of the crust gave way to tension, and just to the east of the new mountains down-faulting and fissures appeared where previously the crust had been squeezed together. Sheets of lava welled up through these fissures and spread out over the sediments that had filled the downfaulted basins. It was one of these sheets of lava that became the huge bulwark of the Hudson River Palisades. Eventually, a decisive line of rifting was established, not far from the line where

the continents had been previously welded together (5). The rift wid-ened to form a sea, which may have looked at first much like the Red Sea today. The continents began to drift apart, and new ocean floor was cre-ated along the mid-ocean rift, by upwelling of basaltic magma from the mantle, a process that is going on today in the middle of the still grow-ing Atlantic [23].

As the new Atlantic Ocean wid-ened, the continental margins tended to subside as they grew more distant from the upwelling ridge at the center of the growing ocean. Vertical adjustments of the crust also took place in response to the trans-

port of eroded materials from the highlands to the continental shelves (6). Little is left today of the high mountain range that once lifted snow-capped peaks where the cities of Boston, New York and Wash-ington sprawl today. If you stand on one of the many outcrops of meta-morphic rock in New York City's Central Park, you are standing on the roots of that ancient range. And in the rugged White Mountains of New Hampshire the "Old Man of the Mountains" cracks a stony smile, pleased to be released at last from his prison of granite in the heart of a mountain.

20 Building the Continents

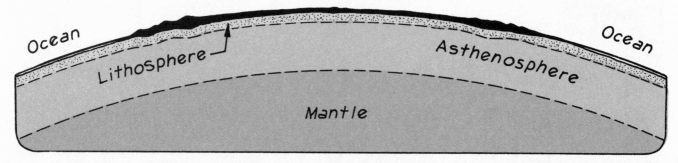

Continent

Ocean

Ocean

Lithosphere

Asthenosphere

Mantle

Alfred Wegener's 1915 theory of drifting continents was dismissed by one of his critics as "a beautiful dream, the dream of a great poet." Poetry, implied the critic, was not science. Before Wegener's dream could become science, it was necessary to see the continents in a new way. It is often the great dreamers and poets who teach scientists how to see.

Wegener's critics believed the continents were the highest standing surface areas of a rock-solid earth, rigid to the bottom of the mantle. We know now that the continents are thin slabs of lightweight rock, hardly thicker on the real earth than the paper skin of a household globe. They lie on the weaker, plastic rock of the asthenosphere. How thin the continents truly are is suggested by the drawing above, a cross-section of North America, where the vertical scale is more or less in proportion with the curvature of the earth. The continental slabs average about twenty to twenty five miles thick, reaching somewhat deeper beneath the mountain ranges. They are composed primarily of granite or metamorphic rocks of a granitelike composition. These rocks are less dense than the basalts of the ocean floors and the crust beneath the continents. Because they are less dense than the basaltic crust of the ocean floor, the continents float higher like blocks of ice floating in water. It is the tendency of these lighter rocks to clump together, like rafts of scum that form on cooking soup, that cre-

ates the ocean basins and areas of dry land. The continents cover about one-third of the earth's surface.

In most places the rocks of the ancient continental slabs (or cratons) are covered with a thin veneer of young sedimentary rock [12]. But in the region of North America centered around Hudson Bay, the slab is exposed in a broad, flat sheet called the shield. Here we are offered a bare glimpse of the long, tortured history of the continent.

The North American shield, like the shields of other continents, is composed of a complex mixture of granitic and metamorphic rocks. The parent rocks of the metamorphics are both volcanic and sedimentary. The metamorphic rocks must have been formed under the conditions of high temperature and pressure that can only be found a mile or more below the surface. Uplift and erosion have exposed the surface we see today. The rocks show signs of intense deformation, as if caught in some terrible squeeze. Most interesting of all are the ages of the rocks of the shield, as determined from the radioactivity of the rocks. The ages of the rocks are summarized on the map at right. In general, the rocks of the shield are younger as one moves outward from Hudson Bay toward the present margins of the continent.

A now generally discredited view holds that most, if not all, of the material of the continents was created by chemical differentiation of the mantle early in the earth's history, perhaps not long after the

planet's formation. But determinations of the ages of shield rocks have led most geologists to the conclusion that the continents have grown by accretion (adding on rims) throughout geologic time, perhaps through the agency of plate convergence. We have seen on the preceding pages how a collision of continents 250 million years ago added new granitic and metamorphic rock to the eastern margin of North America. It is not hard to imagine the deformed belts of ancient granite and metamorphic rock on the Canadian shield as the eroded roots of mountain ranges raised in continental collisions of long ago. The structural trends of the younger shield rocks are more or less concentric with the continental nucleus, consistent with the idea of accretion by continental collisions. The age of the shield rocks suggests that the continents grew in several violent episodes of crustal activity, separated by hundreds of millions of years of relative quiet.

On the scale of the earth, the continents are like thin patches of ice afloat on a pond. That is what the dreamer Wegener helped us to see. Plate motions have moved these "patches of ice" hither and yon, now tearing them apart, now slamming them together. In the violence of collisions the continents have grown in size, fringing themselves with lightweight rocks processed from the material of the upper mantle or reprocessed from the stuff of the continents themselves.

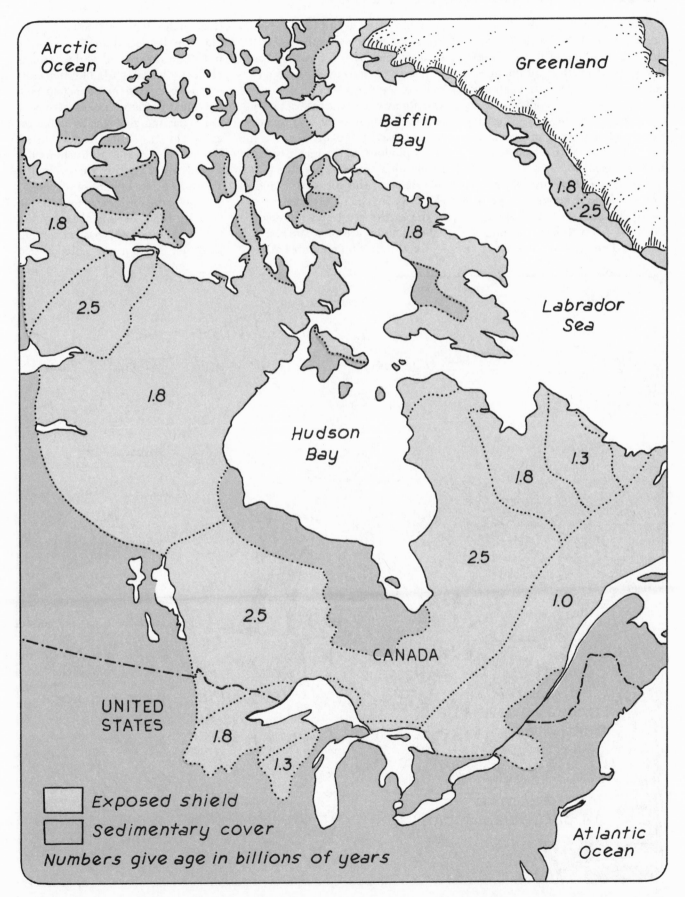

Arctic
Ocean

Greenland

Baffin
Bay

1.8

2.5

Labrador
Sea

1.8

2.5

1.8

1.8

1.3

Hudson
Bay

1.8

2.5

2.5

1.0

CANADA

UNITED
STATES

1.8

1.3

Atlantic
Ocean

Exposed shield
Sedimentary cover
Numbers give age in billions of years

21 The Oldest Rocks

Most geologists agree that the continental crust of the earth has grown through geologic time. It seems likely that in the subduction of oceanic plates beneath continental margins and in the collisions of the continents themselves, material of the upper mantle is processed into the lighter stuff of which the continents are made.

But how old are the oldest continents? How large were they? How did they come to be? These questions do not yet have definite answers. The age of the earth is generally believed to be 4.6 billion years, the same age as the rest of the solar system. This estimate is based primarily on a study of the radioactivity of meteorites and moon rocks. The atoms of certain radioactive elements disintegrate at known rates. By comparing the quantity of the decay products to the quantity of the parent elements it is possible to use the atoms of the rocks as a kind of clock. So far, the oldest known rocks in the earth's crust give radiometric dates of about 3.8 billion years.

Exposures of Precambrian rock (rock more than 600 million years old) can be found in the cores of mountain ranges and at the bottoms of deep canyons [5], but the most extensive formations of ancient rock are found in the sediment-free, generally flat and geologically stable regions of the continents called shields. The map at right shows the shield areas of the earth's crust. These rocks are billions of years old.

One of the current winners in the ancient rock sweepstakes can be found on the west coast of Greenland, not far from the capital of that

Congealed lava

Intruded dike —
2.6 billion years old

3.75 billion years old rock

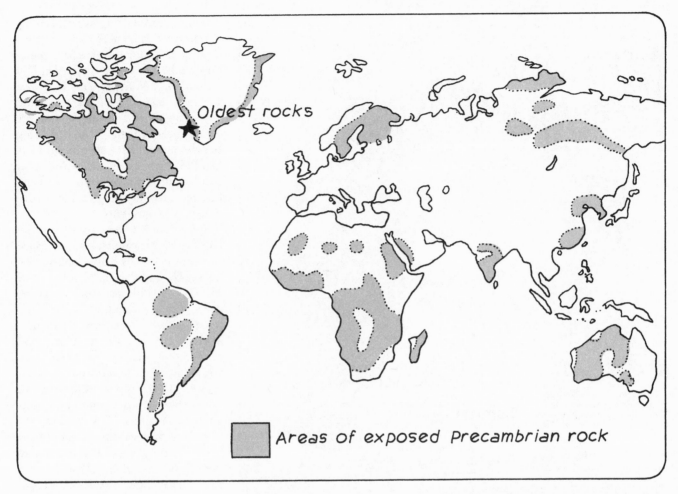

Oldest rocks ★

Areas of exposed Precambrian rock

island, Godthab. The entire Godthab region (see map page 56) consists of very old rocks having complex relationships to one another. They include metamorphosed volcanic and sedimentary rocks, lighter-colored metamorphic rocks of a granitelike composition known as gneisses, and granite. The rocks are highly deformed, and criss-crossed with dikes and seams of younger igneous intrusions. The drawing on the left shows a geologist standing on some very old and highly deformed gneiss near Godthab. The nearby formations are almost equally ancient. These rocks tell a story of being subjected again and again to great temperatures and stress.

But the remarkable thing is that these rocks continue to endure, and that they are so similar in composition and structure to rocks that are being created even today in the forge of plate tectonics. Nowhere on the crust of the earth are there ocean floor rocks more than 200 million years old. Ocean floors are continuously created along the mid-ocean ridges from the upwelling material of the mantle [23], and continuously consumed back into the mantle by subduction at the trenches [52]. No ocean floor has survived for more than a few hundred million years. But the continental rocks are light, and resist being pulled back down into the mantle for reprocessing. They bob like corks above the subduction trenches, they bend and break, but they stay on top. And any new light materials which are differentiated from the heavier stuff of the mantle stay on top with them.

Because the west Greenland rocks contain metamorphosed marine sediments, they tell us that as long as 3.8 billion years ago the earth had an ocean, and in some places a crust that rose above the waves. Apparently, those earliest continents were small, perhaps no more than ten percent of the area of the present continents. Those primitive continents were the nuclei about which the present continents were assembled.

The least known chapter of the earth's history is the interval between the condensation of the planet from the solar nebula 4.6 billion years ago and the formation of the Greenland gneisses 3.8 billion years ago. It may be that it took most of that time for our hot young planet to form a solid crust. Or perhaps still older continental rocks are waiting to be discovered in some remote corner of the world. So far, the crust of the earth is like a book with its first chapter missing.

22 Death on the Ice

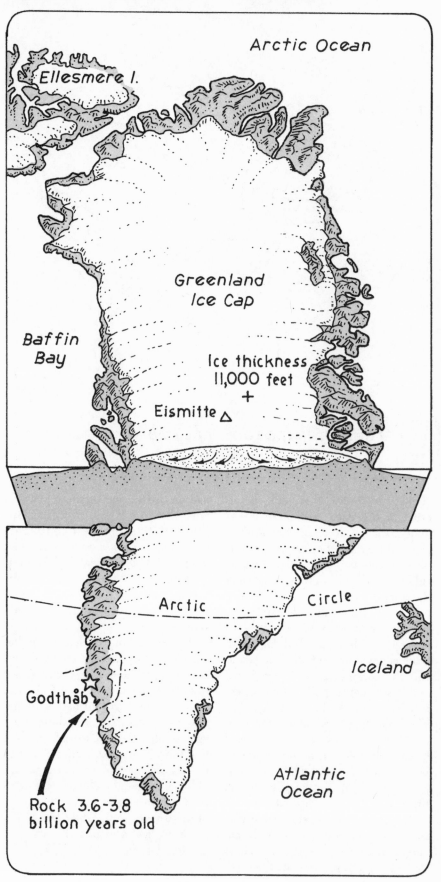

Arctic Ocean

Ellesmere I.

Greenland
Ice Cap

Baffin
Bay

Ice thickness
11,000 feet
+

Eismitte △

Arctic   Circle

Iceland

Godthåb

Rock 3.6-3.8
billion years old

Atlantic
Ocean

Alfred Wegener, the "father of continental drift," had a long and—as we shall see—ultimately tragic association with the ice-capped island of Greenland. Wegener made three scientific journeys to that beautiful and desolate land. It may be that his experience with the rifting, drifting sea ice on the first of those journeys planted the seed in his imagination that would ultimately grow into the first scientifically respectable theory of drifting continents.

Greenland offers an intriguing glimpse of what much of the surface of the earth may have looked like during the great ice ages of the past. The Greenland ice cap covers 80 percent of the island. Only along the coast, where the climate is moderated by the sea, does melting and evaporation manage to keep pace with the accumulation of winter snows. The ice cap is more than two miles thick at its center, but thins toward the coasts. This great weight of ice depresses the earth's crust beneath the glacier. The interior of the island has been depressed in some places below sea level. Mountains ring the coasts. If the ice could be made to suddenly disappear, the island of Greenland would have the appearance of a shallow bowl filled with sea water.

Pressed by its own weight, the Greenland ice cap creeps outward toward the margins of the glacier at a rate of several inches per day. Where the ice reaches the coastal mountains it makes its way over or around these barriers to the sea (see illustration below right). As the ice squeezes through mountain passes it may move with a velocity as great as several feet per hour. In many places along the coast the ice manages to reach the sea, and is there the source of the numerous icebergs that plague shipping in the North Atlantic.

Alfred Wegener returned to Greenland in 1930, as leader of a party of scientists who would spend a year and a half making geological and meterological observations. He was at that time almost fifty years old. His book *The Origins of Continents and Oceans*, in which he had presented evidence for continental drift, had by now been thoroughly debated by earth scientists and just as thoroughly rejected. A key element of the new expedition to Greenland was the establishment of a meteorological station at the center of the ice-cap which would be manned through the cold, dark Arctic winter. The station was named Eismitte, or "mid-ice." The team made three journeys with dog sledges across 200 miles of ice to equip the station. But supplies were still insufficient. After a series of unexpected delays, Wegener set out from the coast in late September with supplementary supplies for the two man team which would winter over at Eismitte. The season was late and the weather terrible. Temperatures had fallen below −50 degrees Fahrenheit. Most of the supply party turned back after a hundred miles, but Wegener and two companions pressed on, arriving at the mid-island station on October 27. The bulk of the supplies they had hoped to bring to Eismitte had been cached or abandoned along the way. There were certainly not enough rations for five men to last through the winter. Wegener celebrated his fiftieth birthday at Eismitte. Then Wegener and his eskimo companion, Rasmus Villumsen, set out in a brave attempt to reach the coast. The two men and their dogs were exhausted. It was a heroic race against death.

Wegener's body was found in the spring, about halfway to the coast. He had died of heart failure, or perhaps of sheer exhaustion, and had been carefully buried by Villumsen. Villumsen attempted to complete the journey alone. His body was never found. The story has another tragic side. Wegener would have been eighty-five years old in 1965, as the new theory of plate tectonics snowballed toward almost universal acceptance. Except for his tragic death on the Greenland ice cap, Wegener might have lived to see his radical, even poetic ideas about drifting continents become the cornerstone of a new geology.

One other objective of Wegener's expedition to Greenland was the measurement of the thickness of the ice cap by echo sounding. The cross section on the map at left can be drawn because of the techniques of seismic sounding pioneered by Wegener and his team.

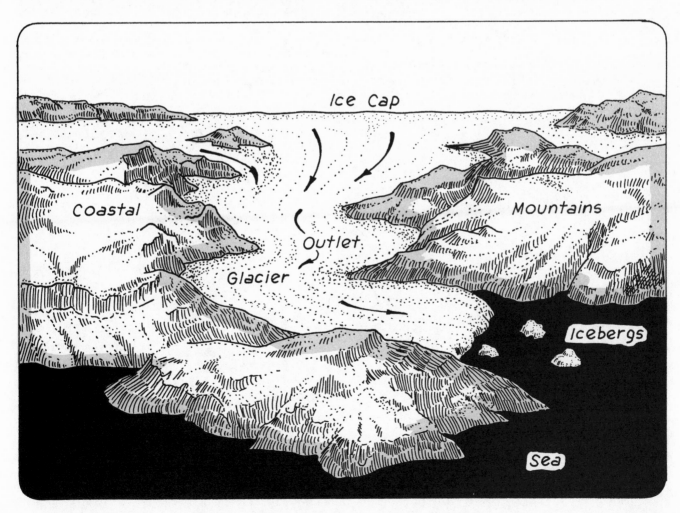

23 Fire and Ice

Not far away from the almost 4 billion year old rocks of the coast of Greenland can be found some of the youngest rocks of the earth's crust. No part of the island of Iceland is more than a few tens of millions of years old. The very oldest rocks of Iceland are in the extreme east and west of that island. The youngest rocks are at the center of the island and lie along a north-south axis. They are only as old as yesterday. Indeed, the central axis of the island is a seething ground of volcanic and geothermal activity. Iceland is still growing—from the inside out.

We could better understand the source of thermal activity in Iceland if we could drain dry the North Atlantic Ocean, as in the drawing below. As the water drained away, first the continental shelves would dry out, and then the continental slopes. At last, when the water level had fallen three or four miles, the deep ocean floor would be revealed. And along the axis of the Atlantic, from a point between the southern horns of Africa and South America all the way to the Arctic Ocean, a sinuous, broken mountain range as high as the Alps would emerge from the waves. The ridge has a double crest. It is sliced down its spine by a deep rift. This is the Mid-Atlantic Ridge and Mid-Atlantic Rift.

The Mid-Atlantic Ridge is the boundary between the North American and South American plates on the west and the European and African plates on the east. It is a divergent boundary, where the plates are

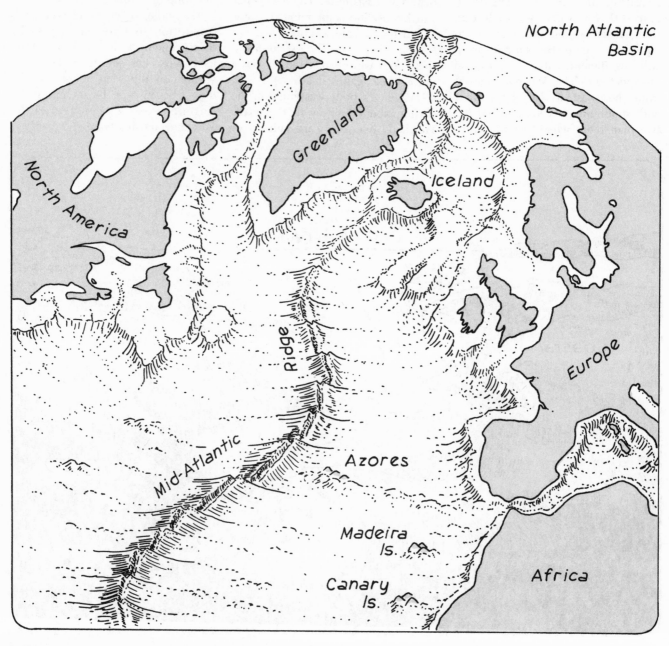

North Atlantic Basin

58

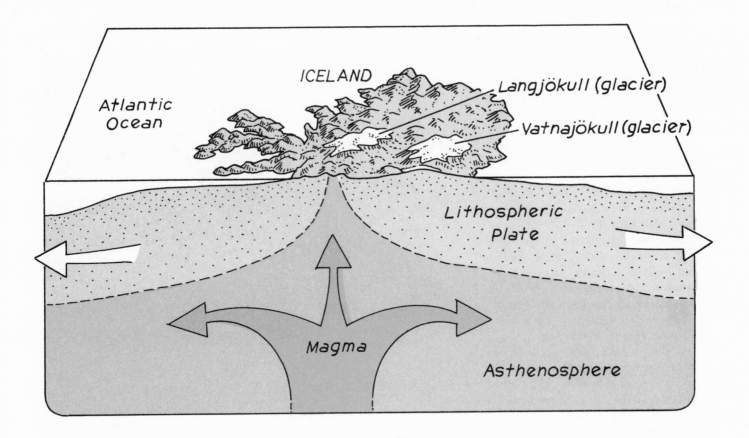

ICELAND

Atlantic Ocean

Langjökull (glacier)

Vatnajökull (glacier)

Lithospheric Plate

Magma

Asthenosphere

moving apart. As the plates diverge, new ocean floor is created by slushy volcanic material rising from the upper mantle to fill the rift. The entire floor of the Atlantic Ocean has been created in this way during the past 180 million years.

Iceland is a particularly active "hot spot" along the ridge. Here, for some reason not yet well understood, a huge rising plume of lava has poured out through the fractured crust to form a broad basaltic plateau. It has been suggested that the impact of a large meteorite, directly on the mid-ocean ridge where the crust is thin and the mantle hot and slushy beneath, may have unleashed the flood of lava. As we shall see later [34], there is contested evidence that the earth was indeed struck by an asteroid at about the time the North Atlantic began to open up. But that such a celestial blow should fall exactly on the spine of an oceanic ridge seems a bit too much of a coincidence.

Iceland is not made of the stuff of continents (the granites and gneisses we saw in Greenland), but is a high-standing block of sea floor. The cracks and fissures of the divergent plate boundary which are elsewhere hidden by the waves are in Iceland made visible. The old Icelandic capital of Thingveliir, for example, lies in a rift valley. Hot water bubbles up from the ground, as it does at Yellowstone. Dikes of lava well up to fill any new fissure in the stretched earth. Near Thingveliir, Iceland is being pulled slowly apart.

Geologists have measured the rate of separation directly, by survey across the rift. The plates are pulled apart at a rate of about a half an inch per year. In the thousand years since Thingveliir was settled by the Norsemen, the valley in which it sits has widened by forty feet. Icelandic landowners who wish to increase their holdings need only wait.

The process by which the island grows is not always a gentle matter of

lava welling up to fill small fissures. Occasionally there is a violent eruption. In 1963, an explosive eruption just off the southern coast of Iceland created the new island of Surtsey [24]. In 1973, a volcano on the island of Heimaey which had been dormant for thousands of years, came dramatically to life (see map page 63). Huge walls of lava poured out of deep rifts in the crust and for a time threatened to completely destroy the town of Vestmannaeyjar, one of the chief fishing ports of Iceland. Many buildings of the town were engulfed by the tide of molten rock and the harbor was nearly sealed off from the sea.

The stretching and tearing of the earth's crust in Iceland, and the upwelling heat of the mantle, make that island a land of fire. Its far northern latitude causes the volcanic mountains to be capped with great glaciers. No other place on earth offers so startling a contrast of fire and ice.

24 Wandering Poles

The earth's magnetic field has proved to be an invaluable source of clues for geologists hoping to reconstruct the past history of the earth's crust. Certain rocks, volcanic lavas among them, contain minerals such as the iron oxide magnetite, and can become weakly magnetized by the earth's magnetic field as they cool from the molten state. Each iron atom acts like a tiny compass needle, and tends to align itself with the local magnetic field. At temperatures above 500 degrees C, the thermal agitation of the atoms jiggles them out of alignment. As the temperature drops, the atoms begin to line up parallel to the earth's field. At about 450 degrees C the atomic alignments are "frozen into place," and the volcanic rock has become permanently magnetized. The rock retains a kind of fossil imprint of the direction of the earth's magnetic field at the time of its solidification.

Consider, for example, the new island of Surtsey, which explosively erupted from the North Atlantic just south of Iceland in 1963 (see map page). Layer by layer the fresh lavas, pouring from a fissure in the sea floor, built up a substantial addition to Iceland's living space. And layer by layer, as the lavas cooled, the iron-bearing mineral grains lined up with the earth's magnetic field. Surtsey is a kind of frozen compass, pointing with every grain of its substance to the present magnetic pole of the earth. The north magnetic pole of the earth is presently among the islands of northern Canada, and Surtsey's fossil magnetism points in that direction.

The study of the paleomagnetism (ancient magnetism) of the earth's crust began in earnest in the 1950s with the development of highly sensitive instruments for measuring the strength of weak magnetic fields. Volcanic rocks can be absolutely dated using radioactivity. It turned out that the paleomagnetism of most rocks of the same age on a given

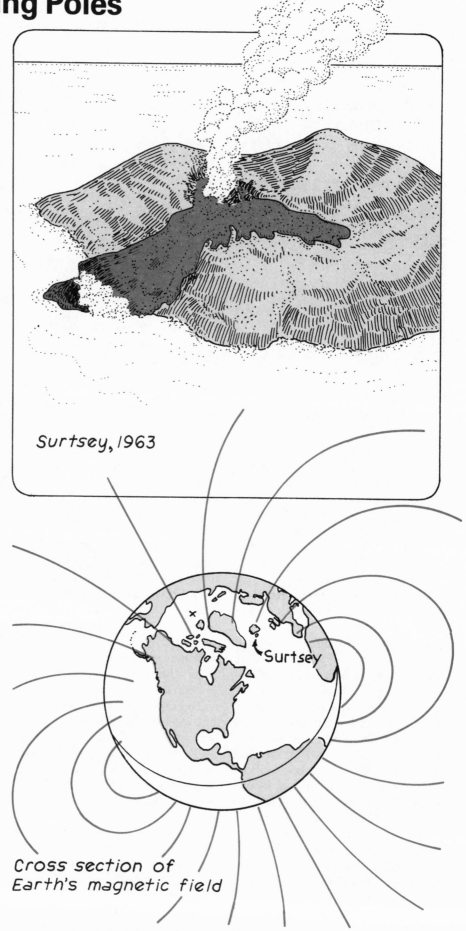

Surtsey, 1963

Cross section of
Earth's magnetic field

continent pointed to roughly the same place on the globe. That place can be assumed to have been the magnetic pole of the earth at the time the rocks were formed. In this way it became possible to reconstruct the positions of the magnetic poles in times past—with respect to the present continents. The positions of the poles have apparently changed. A trace of such positions is called a "path of polar wandering." The map below shows past positions of the north magnetic pole as inferred from the rocks of North America and Eurasia.

Most geologists are inclined to believe that the earth's magnetic poles have never wandered far from the geographic poles defined by the earth's axis of rotation. If this is the case, then it is not the magnetic poles that have wandered, but the continents. This interpretation is supported by the seeming fact that the paleomagnetism of different continents point to different paths of polar wandering (see map below). Before about 100 million years ago, the rocks of North America and Eurasia indicate different positions for the earth's magnetic poles. It seems highly unlikely that the earth had two or more magnetic poles in the past. The most widely held conviction is that the divergent paths of polar wandering reflect the changing orientations of the continents with respect to each other and to the axis of rotation of the earth. As more paleomagnetic data is accumulated, geologists will be able to provide better reconstructions of past configurations of the continents.

The study of paleomagnetism relies on statistical correlations of very rough data and is fraught with practical and theoretical difficulties. For one thing, the source of the earth's magnetic field is itself poorly understood. For another, the use of rocks as fossil compasses relies on the assumption that the rocks have not been tipped or displaced relative to the continent since their formation. But despite these and other doubts, geologists are becoming increasingly comfortable with the story of drifting continents as recorded in the magnetism of the ancient rocks.

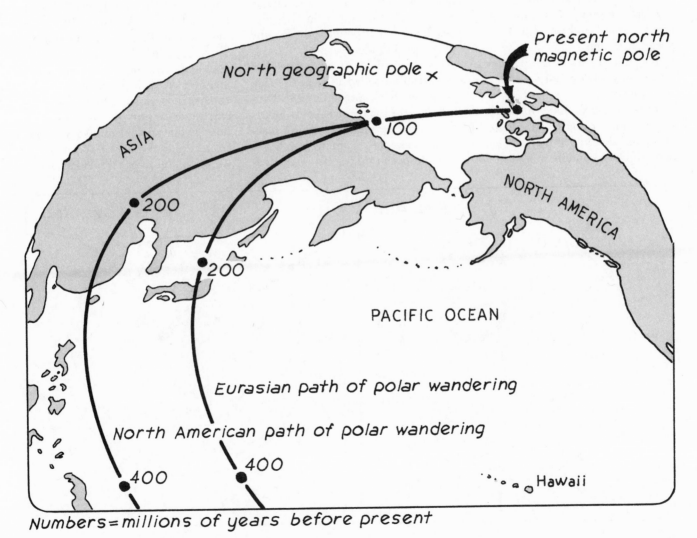

25 Magnetic Stripes

There is not yet a satisfactory theory to account for the earth's magnetic field. It seems clear that the field has its origin in electrical currents that flow deep in the earth's liquid core. What "stirs" the core and keeps the currents flowing remains a mystery. A surprising and unexpected feature of the earth's magnetic field was recognized during the 1950s and 1960s: the polarity of the magnetic field has reversed many times in the past! Fossil magnetism in crustal rocks record not only past positions of the magnetic poles, but also their polarity. At least ten times in the past 4 million years the north and south magnetic poles have flipped, just as the field of a bar magnet would reverse if the bar were twisted 180 degrees. After such reversals, which occur at irregular and unpredictable intervals, a compass needle would swing around and point in the opposite direction.

A careful study of the paleomagnetism of sea floor sediments deposited over millions of years has shown that reversals take some thousands of years to be accomplished. Grains of sediment slowly drifting to the sea floor, if they contain iron, can be-

come aligned like compass needles as they settle onto the bottom. The effect is weak, but there will be an average alignment of grains as the sediments build up. Sediments can therefore record the strength and polarity of the earth's magnetic field at the time of deposition. During a reversal, the sea floor sediments show, the earth's field slowly decreases in strength until it reaches zero, and then rebuilds in the opposite direction.

Reversals of the earth's magnetic field are undoubtedly related in some way to reversals in the direction of flow of electrical currents in the core, but as yet no one knows how or why these flip-flops occur. Polarity reversals appear to have been a regular feature of the earth's field throughout its history, but an accurate chronology of the flip-flops has been constructed only for the past 4 or 5 million years.

Magnetic field reversals provide a remarkable confirmation of the theory of sea floor spreading [23]. During the late 1950s, marine geologists discovered strange patterns of magnetic "anomalies" running parallel to the mid-ocean ridges. An anomaly is

a region where the strength of the magnetic field measured at the surface of the earth is a little bit stronger (positive anomaly) or weaker (negative anomaly) than the average field. The drawing at right shows the pattern of anomalies revealed by a magnetic survey in the region of the Mid-Atlantic Ridge south of Iceland. Regions of stronger and weaker field are shown in black and color respectively.

In 1963 British geologists Fred Vine and Drummond Matthews offered an imaginative, and now widely accepted, explanation for the anomalies. They began with the theory of sea floor spreading, itself very new in 1963. According to that theory, oceans grow by the continuous production of new oceanic crust from magma rising from the mantle along the mid-ocean ridges. Plate movement carries the fresh sea floor away from the ridge. Vine and Matthews assumed that the fresh basalts are magnetized with the polarity of the earth's field at the time they are extruded. When the earth's field flips, so does the magnetism which is "frozen" into the growing ocean floor. As we move across the ridge today,

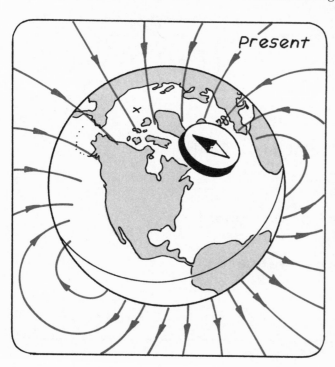

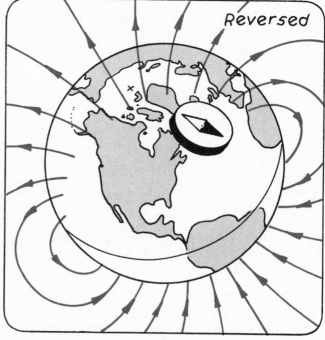

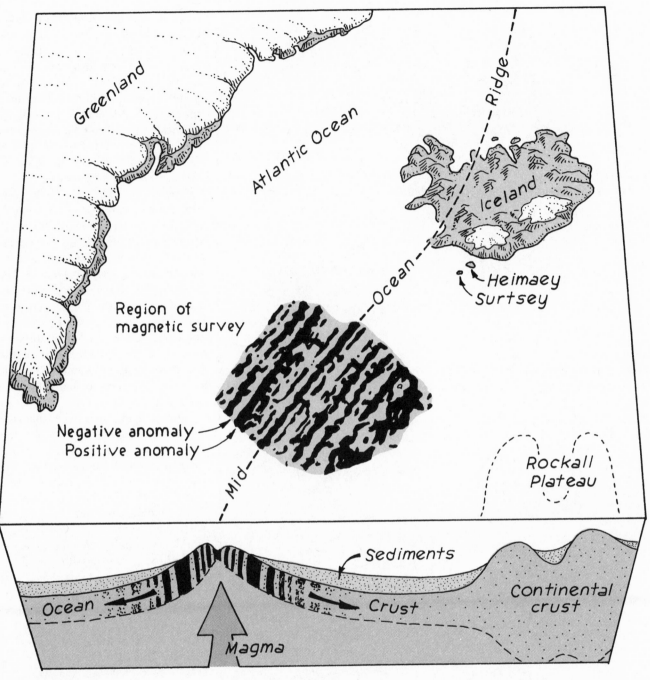

the fossil magnetism of the ocean crust adds or subtracts its own weak contribution to the present field of the earth. Positive anomalies correspond to places where the crust beneath the sea is magnetized with the same polarity as presently. Crust beneath a negative anomaly was formed when the earth's field was opposite to its present polarity. The spreading sea floor has acted as a kind of tape recorder, recording in stereo the field reversals of the past. Conversely, the field reversals, recorded worldwide on ocean floors, preserves a vivid record of the direction and rate of sea floor spreading.

During reversals there may be a period of thousands of years when the magnetic field is nearly zero. There has been much speculation regarding the effect of cosmic rays and high energy solar particles on an earth unprotected by its magnetic field. The field tends to deflect and trap charged particles and keep them from reaching the surface except in polar latitudes. Some scientists believe that the increased bombardment of the earth by high energy particles during periods of field reversal may have caused extinctions and transformations of living species.

26 A Lost Continent

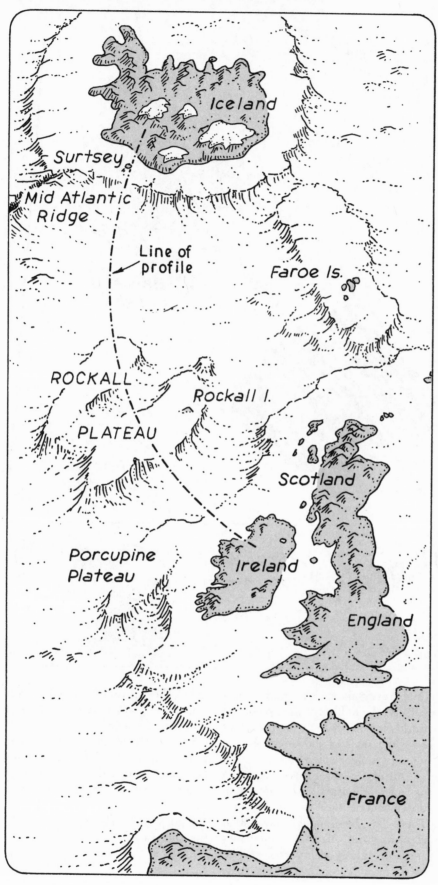

Iceland is a mound of "ocean floor" that just happens to stand above sea level. Not far to the southeast of Iceland there is a little slab of "continent" with a surface almost entirely under water. The Rockall Plateau, a submerged table of rock in the North Atlantic, appears to be a fragment of continental crust that broke off from Europe about 60 million years ago and sank beneath the waves.

The Rockall Plateau is as large in area as New England. It stands several miles above the ocean floor, but its surface is covered by the waves. Only tiny Rockall Islet sticks a tip of rock above the watery element. The islet offers a lucky opportunity for geologists to study the structure of the plateau from dry land. A visit to the islet confirms that its material is typical of the granitic stuff of the continents, rather than the basaltic rocks of the ocean floors (and of Iceland).

The primary evidence that the Rockall Plateau is continental rather than oceanic came through seismic investigations. Seismic waves (from earthquakes or from artificial explosions) travel with different velocities through different kinds of rock. In particular, such waves move more rapidly through continental crust than through the basaltic sea floor. A measurement of the velocities of reflected shock waves from artificially set seismic impulses confirmed the granitic nature of the plateau. The seismic studies also revealed that the Rockall Plateau has deep granite roots that reach far below the ocean crust. The submerged granite bank has a thickness identical to that of the nearby continental crust in Northern Ireland (see profile top of next page).

But still other evidence confirms that the Rockall Plateau is a drowned mini-continent rather than raised sea floor. The deep-sea drilling ship *Glomar Challenger* [48] sank several bore holes on the Rockall Plateau. The vessel has the ability to bring up

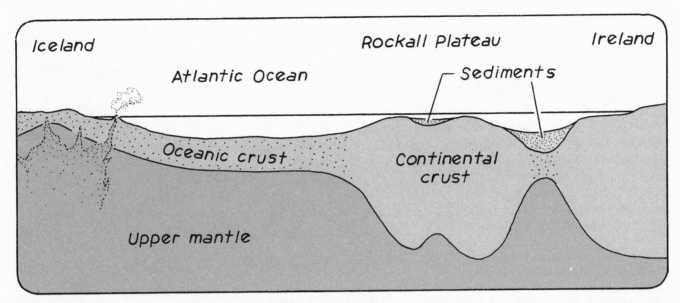

Iceland   Rockall Plateau   Ireland

Atlantic Ocean   ⌐ Sediments

Oceanic crust   Continental crust

Upper mantle

from great depths a long vertical sample of sea floor sediments, called a core. The nature of the fossil protozoa and microscopic marine plants in the sediments are indicators of the depth of the water in which the sediments were deposited. The cores revealed that the Rockall Plateau had once been dry land, and that it sank in three separate episodes of subsidence followed by relatively stable intervals (see drawing below).

A last indication of the continental character of the Rockall Plateau, which can be seen in the drawing at left, is the nice jigsaw puzzle fit of the plateau with the adjacent continental slope.

Sixty million years ago, when the Rockall continental fragment broke away from Ireland, the Atlantic was only about a third as wide as it is today. North America, Greenland and Europe were moving away from one another, with some twisting and turning and squeezing. Drifting continents on a spherical planet are subject to a complex system of tension and compression. It was tension in the crust that broke the Rockall Plateau away from the mainland. We shall see other evidence of this tensional stress in Northern Ireland and Scotland.

Several other submerged minicontinents have been identified at the Seychelles Bank in the Indian Ocean and the Ontong-Java Plateau in the Pacific. The existence of a broad slab of submerged continent in the middle of the Atlantic evokes the legend of the Lost Continent of Atlantis. But it is hardly possible that the Rockall Plateau, which sank beneath the waves over 50 million years ago, could be the source of a legend that first appeared in the writings of Plato, only a few thousand years ago. Besides, there is a much more likely source for the legend of Lost Atlantis—as we shall see when our geological journey reaches the Mediterranean.

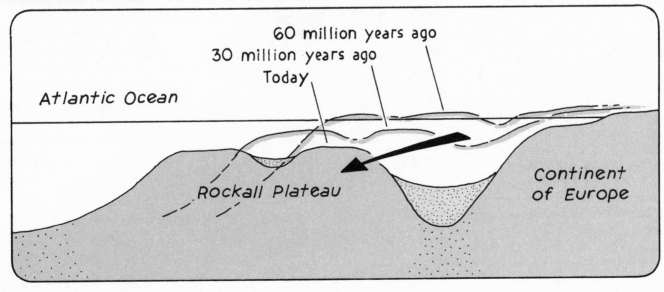

60 million years ago
30 million years ago
Today

Atlantic Ocean

Rockall Plateau   Continent of Europe

27 Giant's Causeway

At the same time the Rockall undersea plateau was torn from the flank of Europe, the whole northwestern corner of that continent was subjected to a terrible tension. This tension had its origin in the shift of plates that accompanied the opening of the north Atlantic Ocean. It was not only between the Rockall fragment and the mainland that a fracture opened up. Elsewhere on the mainland, although less decisively, the crust was pulled asunder. Lavas welled up through these cracks in the earth's eggshell-thin crust to form, in northern Ireland and western Scotland, the kind of extensive basaltic plateaus we saw in the Pacific northwest [8]. In some places these sheets of lava were a half mile thick. Across Scotland and northern Ireland fissures filled with lava to form dikes that radiate across the landscape. Even a cursory glance at a geological map of Britian shows that the northwestern corner of the island has been wracked by powerful violence. The area is quiet today, a place of green glens and rolling highlands, but 50 or 60 million years ago it was a land of fire. And nowhere has the geological violence of those times left behind a more curious and spectacular legacy than at a place on the Irish coast called Giant's Causeway.

Ireland is a land of giants. Its folklore is rich in legends of superheros of the misty past who could lift and hurl huge stones and span broad glens with a single stride. And how else but as the work of giants would our ancestors in those mythy times explain the great piles of polygonal stepping stones that project from the green coast of County Antrim into the emerald sea? Even the modern visitor, versed in the skeptical ways of science, is likely to see in the astonishing geometrical regularity of these stone columns the work of the hand of giants.

Basaltic lavas, unlike the thicker tarlike lavas that erupt from cone-

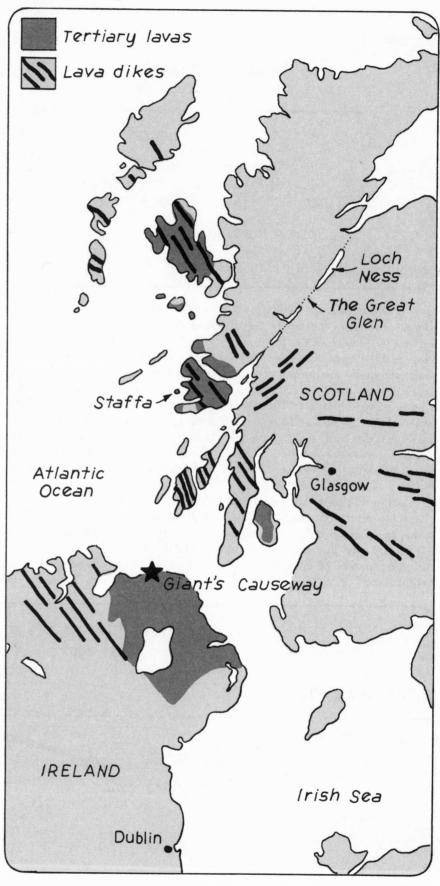

The Giant's Causeway

shaped volcanoes, have a low viscosity. Pouring out from fractures in the earth's crust, they readily flow across the surface to form the broad flat sheets of black rock that we see in northern Ireland and western Scotland. As the lavas cease to flow and begin to cool, solidification begins at the top and bottom surfaces of the sheet where the molten rock is in contact with the cool air and cool base. As the lava begins to crystallize, it contracts, and tensional stresses build up inside the sheet. These stresses tend to lie in planes parallel to the surfaces of the lava sheet. Vertical fractures form. In uniform flat sheets of lava these frac-

tures are regularly spaced, and the solidifying basalt breaks up into polygonal cylinders. The vertical columns are often hexagonal in cross section, like bunches of pencils standing on end. At the Giant's Causeway these cylinders have been exposed by erosion for the appreciation of tourists and the delight of children. As in snowflakes and honeycombs and the seeds of pomegranates, marvelous mathematical form here manifests itself as part of nature's architecture. "Nature plays," said the Renaissance astronomer Johannes Kepler, as he considered the snowflake's hexagon. At the Giant's Causeway we see nature at

play!

There is an equally spectacular, although less accessible, display of these vertical columns at Fingal's Cave on the island of Staffa off the coast of Scotland, a place that has been made famous through the music of Mendelssohn. The name of the cave commemorates a mythy giant of the Irish past, and it is easy to imagine that Fingal once made his way across the sea from Ireland to Scotland on stepping stones that have now become mostly submerged beneath the waves. At Giant's Causeway the stress that tears crustal plates has given rise to the stuff of dreams.

28 The Monster's Lair

The Great Glen

Anyone who looks at a map of Britain, especially a map that shows topography, cannot fail to take note of what looks like the slash of a knife across the Scottish Highlands. This is the Great Glen, a deep, straight valley that cuts diagonally across the island from Inverness to Fort William. At each end of the Glen long arms of the sea intrude upon the land, the Moray Firth in the northeast and the Firth of Lorne in the southwest. Along the floor of the Glen are a string of lakes, including Loch Ness, the reputed home of the famous monster. To either side of the Glen are some of the highest mountains in Britain.

Geologists have long recognized that the Great Glen lies along a transform fault, a break in the crust of the earth where one part of the crust has slid laterally with respect to another. Another transform fault we have visited is the San Andreas Fault in California, and there are many similarities between the two fault systems. The fault zone is a belt of crushed, sheared and metamorphosed rock up to a mile wide. The apparent gash of the Great Glen has resulted from erosion of the weakened rocks along the fault zone. What has been recognized more recently is that the Great Glen Fault is an extension of the Cabot Fault, another crack in the crust that extends from Boston, Massachusetts northward along the coast of New England to Newfoundland. 350 million years ago when there was no Atlantic Ocean (see map, page 45) one could

have walked along the fault trace from Boston to Inverness without getting one's feet wet. The recognition that the two faults match up when the Atlantic continental jigsaw puzzle is reassembled helped confirm the theory of drifting continents and a growing Atlantic. The continuity of the faults is just one of several geological structures that are continuous across the Atlantic. The pieces of the jigsaw fit snugly together—and the pictures match!

During the time the Great Glen Fault was active, the two parts of Scotland divided by the fault moved at least sixty-five miles with respect to one another, as illustrated by the maps on these pages. (Bear in mind, of course, that the first of these maps cannot give a realistic picture of the

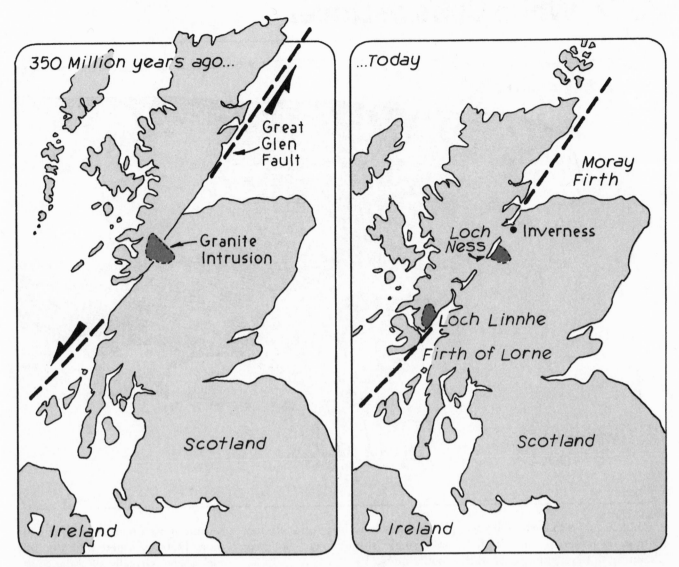

350 Million years ago...

Great Glen Fault

Granite Intrusion

Scotland

Ireland

...Today

Moray Firth

Loch Ness • Inverness

Loch Linnhe

Firth of Lorne

Scotland

Ireland

topography and coastlines of 350 million years ago. Only the relative position of the ancient rocks of Scotland should be inferred from the map.) The most impressive pieces of evidence for displacement along the fault are the ancient intrusions of granite exposed on the south side of Loch Ness and the north side of Loch Linnhe. The intrusions are sixty-five miles apart and are apparently of about the same age. Each intrusion is roughly semicircular in shape and is truncated by the fault. The conclusion seems inescapable that they are the two halves of a single granitic intrusion that predates slippage along the fault.

The real action along the Great Glen Fault took place hundreds of millions of years ago. This was the time when the Appalachian Mountains were being pushed up in North America and the Caledonian ranges were thrust up in Great Britain and Scandinavia. We now know that these great events were the result of continental collision, what we earlier called "the Big Crunch" [19]. The Glen is relatively quiet today. Earthquakes still occur along the fault, but most are minor. The last strong shock occurred at Inverness in 1901.

Those people who believe that a monster lurks in the inky depths of Loch Ness often suggest that the creature is a marine dinosaur, a plesiosaur perhaps, that has somehow survived in the waters of the lake though elsewhere extinct for 65 million years. According to this delightful but totally improbable the-

ory, a wrench of the fault cut off an arm of the sea and trapped Nessie (as the monster is affectionately called) in her lair. Believers recall the case of the coelacanth, a primitive fish thought to have been extinct for 70 million years before one was discovered in the Indian Ocean in 1938. If a coelacanth survived in the ocean depths, why not a plesiosaur in the 900 foot depths of Loch Ness? Why not, indeed! But few scientists take the claims of the Nessie watchers seriously, and none believe the monster might be a trapped dinosaur. For one thing, only ten thousand years ago northern Europe was in the throes of an ice age and the Great Glen was filled with ice, not water. There was no dark lake for Nessie to hide in then!

29 White Cliffs of Dover

White Cliffs of Dover

For millions of visitors who approach Britain from across the English Channel, a first glimpse of that storied island consists of the high coastal cliffs near the ferry port of Dover. The cliffs are remarkable for their steepness and brilliant whiteness. They rise hundreds of feet abruptly from the sea and are composed entirely of chalk.

The sheer verticality of the cliffs at Dover is due to the erosional weakness of the chalk. Chalk is a crumbly sort of rock and fails before the fierce wind-driven waves that funnel up the English Channel from the stormy Atlantic Ocean and North Sea. The cliffs are not protected by a base of hard rocky rubble. The waves eat readily into the feet of the cliffs and the overhanging chalk falls into the sea where it is quickly devoured by the waves. In some places the cliffs have been cut back as much as forty to ninety feet in a single stormy night, in great toothy bites.

The cliffs at Dover and other locations along the channel coast of Britain and France are only the most conspicuous outcrops of a thick stratum of chalk which underlies all of southeastern England and much of northern France and Belgium. The bed varies in thickness, but in places is as much as 600 feet thick. Chalk is made up entirely of fossils, the microscopic calcareous skeletons of single-celled organisms that lived a planktonic existence in ancient seas. A common variety of these fossils is the coccolith, illustrated on this page. Coccoliths are tiny hard rossette shells secreted by algae around their soft spherical bodies. At death, the shells drift to the sea floor, where over millions of years (if undisturbed) they can build up thick sedimentary deposits that are ultimately transformed into the soft rock chalk. Virtually all of the continental chalk in the world was deposited in this way during the Cretaceous era, about 100 million years ago. In fact, that geological era takes its name from the thick chalk deposits near the English Channel; the Latin *creta* means "chalk."

The presence of the thick chalk stratum tells us something about conditions on the crust of the earth 100 million years ago. In southeastern England the sedimentary strata (including the buried chalk) have been gently folded, as illustrated at right. The folding was a distant ripple of the thrust which further south threw up the Alps [33]. In some places the uplifted strata have been eroded away, exposing the chalk. In other places (near London, for example) recent sediments have

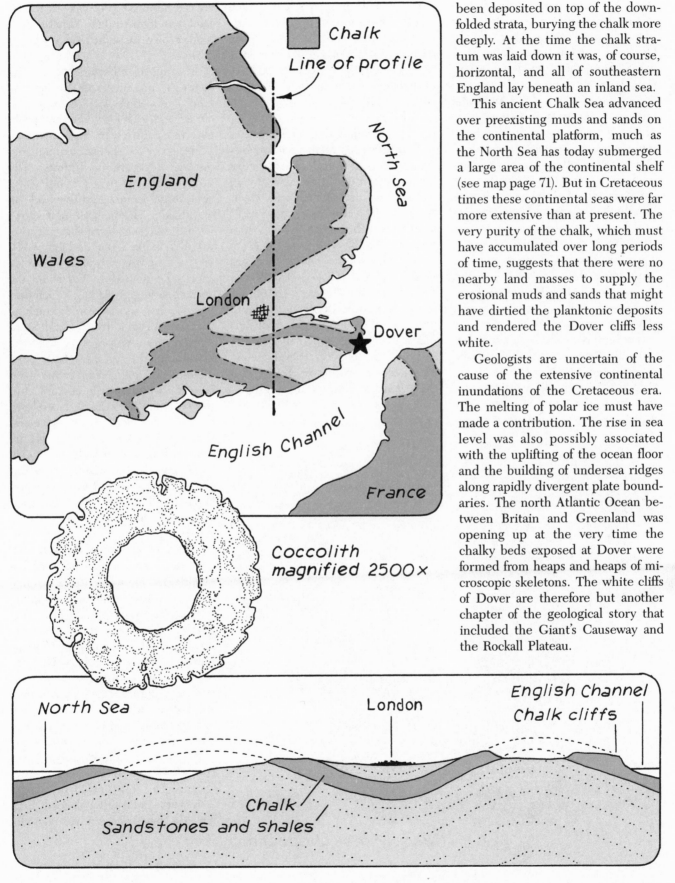

Chalk

Line of profile

North Sea

England

Wales

London

Dover

English Channel

France

Coccolith
magnified 2500×

North Sea

London

English Channel
Chalk cliffs

Chalk

Sandstones and shales

been deposited on top of the down-folded strata, burying the chalk more deeply. At the time the chalk stratum was laid down it was, of course, horizontal, and all of southeastern England lay beneath an inland sea.

This ancient Chalk Sea advanced over preexisting muds and sands on the continental platform, much as the North Sea has today submerged a large area of the continental shelf (see map page 71). But in Cretaceous times these continental seas were far more extensive than at present. The very purity of the chalk, which must have accumulated over long periods of time, suggests that there were no nearby land masses to supply the erosional muds and sands that might have dirtied the planktonic deposits and rendered the Dover cliffs less white.

Geologists are uncertain of the cause of the extensive continental inundations of the Cretaceous era. The melting of polar ice must have made a contribution. The rise in sea level was also possibly associated with the uplifting of the ocean floor and the building of undersea ridges along rapidly divergent plate boundaries. The north Atlantic Ocean between Britain and Greenland was opening up at the very time the chalky beds exposed at Dover were formed from heaps and heaps of microscopic skeletons. The white cliffs of Dover are therefore but another chapter of the geological story that included the Giant's Causeway and the Rockall Plateau.

30 Land Bridge to Britain

Britons have long taken satisfaction in the sea moat that separates their island from the continent of Europe. On more than one occasion the English Channel has provided the barrier that protected the "sceptered isle" from invasion by continental forces. Even today, Britons have mixed feelings about proposals to build a tunnel beneath the strait at Dover, connecting southeast England to France. The twenty-mile wide stretch of water continues to have both strategic and symbolic significance to a people who take a stubborn pride in their insularity.

But it has not always been so. Throughout the past several millions of years the land of Britain has been on many occasions a peninsular arm of the continent. Unlike the mostly submerged Rockall Plateau, which is separated from continental Europe by a narrow stretch of true sea floor, Britain is geologically speaking well and truly a part of Europe. The islands are merely the highest stand-ing parts of a continental shelf which is elsewhere broadly submerged. And the water gap that holds those islands aloof from Europe has long been vulnerable to a rise in the level of the land, a drop in the level of the sea, or both.

Naturalists who seek to relate present and past flora and fauna of the British Isles to the natural history of continental Europe are keenly interested in the schedule of the rise and fall of land and water. Archeologists attempting to reconstruct the prehistory of human visitations to the isles are also eager to know when and for how long a land bridge existed between Britain and the continent. During the period from one hundred to forty thousand years ago, while Neanderthal humans roamed Europe, the bridge opened and closed several times, alternately providing easy access to the islands and stranding migrants. Stone tools, such as those illustrated on this page, give ample evidence of an early human presence in the islands, evidence that Neanderthals made ready use of the bridge when it existed.

The most important controlling factor in opening and closing the land connection with Europe was climate. At least four times and probably more during the past 4 million years, great ice sheets have built up on the northern continents. The huge volume of water stored up in continental ice lowered the sea level worldwide. The map at right shows what the present European coastline would look like if sea level fell by 300 feet, a drop not untypical of the ice ages. The map does not show the corresponding glaciation, which during the ice ages covered much of northern Britain, Ireland and Scandinavia (see next pages). Life at the margins of the ice must have been difficult for our Neanderthal (and later Cro-Magnon) ancestors, but they seemed to have lived and hunted very near to the glaciers, wandering back and forth across the tundra lowlands which are today submerged by the waters of the English Channel and North Sea, perhaps travelling along the conjoined waters of the Thames and Rhine Rivers.

Not only the level of the water but also the level of the floor of the channel has helped control the opening and closing of the land bridge to Britain. The great weight of ice age glaciers depressed the underlying crust, and time was required for the land to rebound when the ice melted. The continental shelf may still be rising from the depression of the last ice age. Quite independent of the comings and goings of the ice, stresses in the shifting continental plates have also helped raise or lower the elevation of the continental shelves. The equation of land and sea at Dover is delicately balanced. Britain may someday again loose its salty moat.

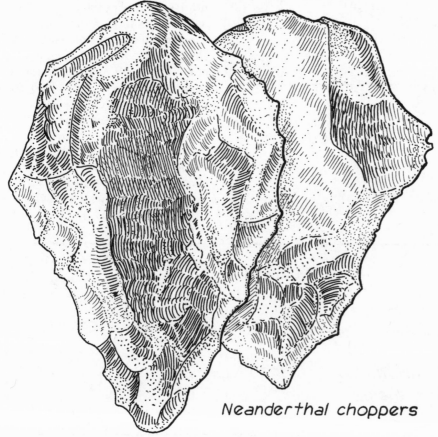

Neanderthal choppers

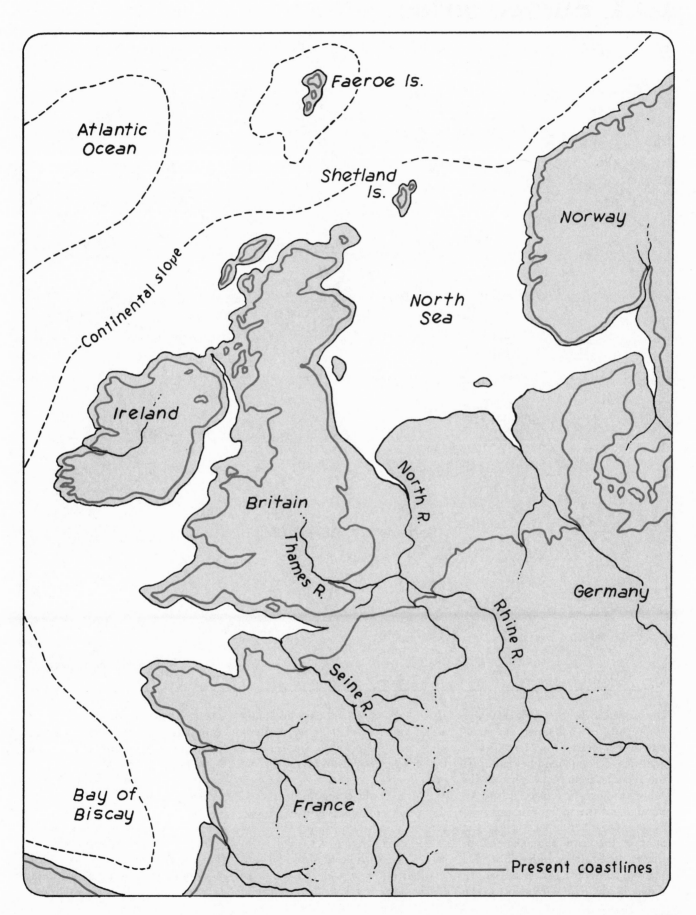

31 Europe on Ice

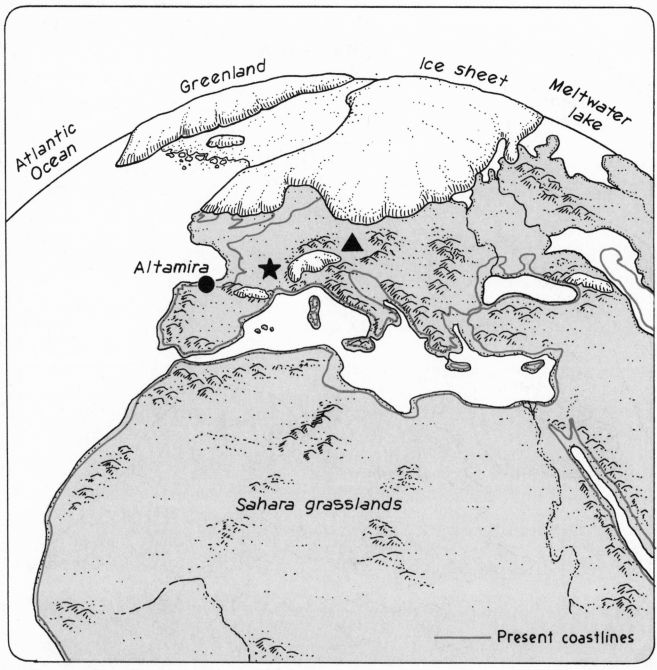

One day in 1879, the young daughter of the Spanish nobleman and amateur archeologist, Don Marcelino de Sautuola, accompanied her father on the reconnaissance of a cave on his summer estate at a place called Altamira on the northern coast of Spain. While her father dug into the floor of the cave searching for artifacts, the girl wandered into a nearby low-ceilinged chamber. In the flickering light of her lantern she saw a herd of red animals streaming across the roof of the cavern. Running back to her father, she shouted the news of her astonishing discovery.

The scientific community greeted the Don's announcement of ice age art with almost total skepticism. Few people of his time were willing to accept these delicate, graceful and brilliantly executed drawings of bison, horses, a wolf, deer and wild boars as the work of prehistoric "cavemen." The drawings, concluded the experts, were probably no more than twenty years old.

But by the end of the century, similar examples of Ice Age cave art had been discovered at other sites in France and northern Spain, and the opinion of the experts began to change. The most extensive and famous of these galleries of prehistoric art, the cave at Lascaux in France, was discovered in 1940. By that time scientists and art scholars had long since acknowledged the authenticity of the little girl's discovery. The conclusion was inescapable. At the very time when much of Europe lay burdened by a thick blanket of ice, the men and women who lived along the bitterly cold margins of the great glaciers achieved a magnificent flowering of creativity and art.

The last ice age in Europe was contemporaneous with the most recent glaciation of North America [14]. From about seventy to ten thousand years ago, glaciers spreading from centers of accumulation in Scandinavia and Scotland pushed down to cover parts of Ireland, Britain, Germany, Poland and the Soviet Union. Smaller glaciers spread out from the Alps, the Pyrenees and the Caucasus mountains. All of Switzerland and parts of her neighbors lay beneath the Alpine ice sheet. The map at left shows the general extent of the ice sheets and takes note of the drop in sea level that accompanied the piling up of so much water upon the continents.

Along the margins of these great ice sheets there was treeless tundra. Herds of reindeer and mammoths grazed on the heather and other low-growing plants of the boggy soil. Life must have been hard for the men and women who hunted these animals, ate the flesh, and warmed themselves in clothing made from the skins. But the very harshness of the conditions of life seems to have sparked a remarkable inventiveness.

The oldest known written records date from this time, notations on bone which probably recorded the phases of the moon. The caves at Altamira, Lascaux and elsewhere were brilliantly decorated, most often with figures of the animals of the hunt, but sometimes evidencing the sheer joy of making images. Delicate figures associated with nature worship were sculpted, such as the clay Venus illustrated below. Toolmaking flourished and the use of the needle became widespread. When the ice began to retreat about ten thousand years ago, it had bequeathed to humankind a rich cultural heritage.

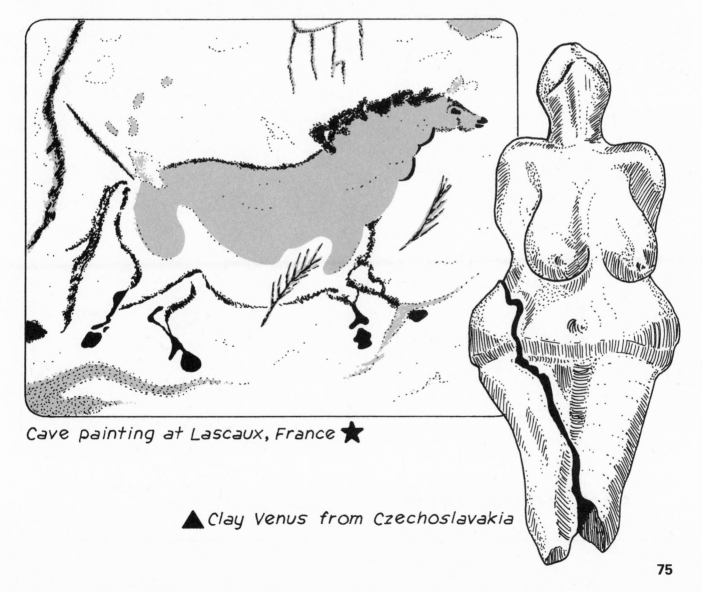

Cave painting at Lascaux, France ★

▲ Clay Venus from Czechoslavakia

75

32 The Tethys Sea

The next few stops on our around the world geological tour will take us to the Mediterranean Basin. We will approach the basin across the mighty barrier of the Alps, and we will stop to admire that splendid range. But if we are to understand the Mediterranean and the lofty mountains that guard its northern rim, we shall first have to take a journey into the past and study the arrangement of the continents as they were disposed hundreds of millions of years ago. We shall see that the Mediterranean is the last remnant of an arm of a much greater body of water, the Tethys Sea, that once divided Europe from Africa.

Two hundred million years ago all of the present continents were joined in a single landmass which geologists call Pangaea ("all-earth"). On the east, a great wedge-shaped arm of the universal ocean intruded deeply into the supercontinent, with its vertex near the present Strait of Gibraltar (see cross on maps below). This now vanished body of water takes its name from *Tethys*, the wife of *Oceanus* in Greek mythology and the mother of seas.

By about 140 million years ago (see map below) rifting of Pangaea had begun to open up the present north Atlantic Ocean. As Africa and North America moved apart the southern continents pivoted about Gibraltar and the Tethys Sea began to narrow. At about this time the slowly widening Atlantic Ocean probably connected with the western end of the Tethys Sea to create an around-the-world low latitude seaway, separating the northern and southern continents. Geologists call the northern landmass Laurasia, after the Laurentian geological province of North America and Eurasia, and the southern landmass Gondwanaland, after the Gondwana geological formation in India. As the two landmasses parted company, evolution proceeded along rather different paths. About 140 million years ago the Laurasian and Gondwana fossil records began to diverge.

Along the subsiding margins of the new earth-girding seaway, sedimentary basins were created (dark triangles on map below). Organic

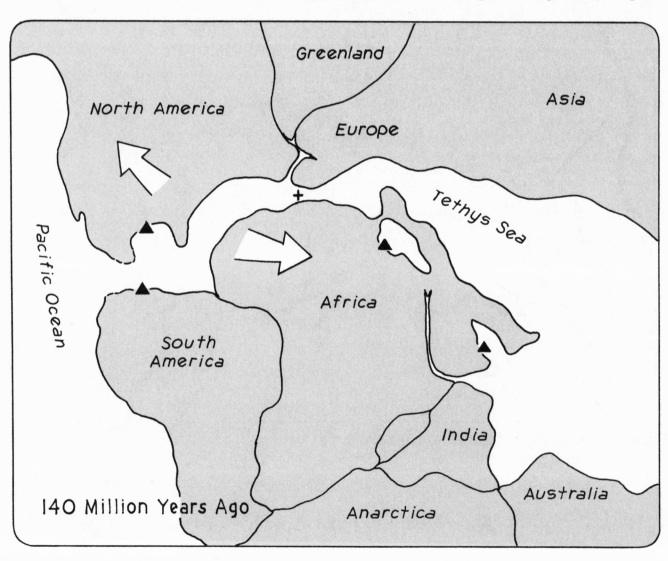

140 Million Years Ago

matter collected in these basins. Much of this organic matter consisted of planktonic microorganisms. Conditions of climate, ocean currents and tectonic activity at that time seem to have given rise to an unusually great proliferation of planktonic organisms. The thick organic deposits in the Cretaceous basins of the Tethys seaway are the probable source of the rich petroleum reserves of the Gulf of Mexico, Venezuela, Libya and the Persian Gulf [46].

The map below shows the arrangement of the continents 60 million years ago. South America has split away from Africa along a continuous rift that runs down the entire length of the Atlantic. In the north, the Atlantic first pushed an arm between Greenland and North America. Then Greenland began to separate from Europe, instigating the sequence of tectonic events in the northern British Isles that have been described on preceding pages [26,28]. In the south, North and South America moved independently westward and were not yet connected at Panama [56]. Meanwhile, Africa continued to pivot on a point near Gibraltar, closing up the Tethys Sea and pushing up the mountains that line the underside of Europe. The Italian peninsula was jammed into Europe with a force that piled up the Alps. The Dolomite Alps in northeastern Italy were carved from Tethyan reefs which were caught in the crunch and heaved onto the land. Greece and the Balkans are apparently continental fragments which were transferred from Africa to Europe. Cyprus is probably a bit of Tethyan oceanic ridge which was squeezed and elevated above the level of the sea. At Gibraltar, the pressure of Africa against Europe pushed up limestones which had been laid down on the floor of the Tethys seaway—and formed the ancestral uplift of the famous Rock of Gibraltar. Antarctica drifted south and Australia moved east, toward their present locations on the globe. On the map you will note India adrift and heading north. Its fate will be discussed in a later essay [47].

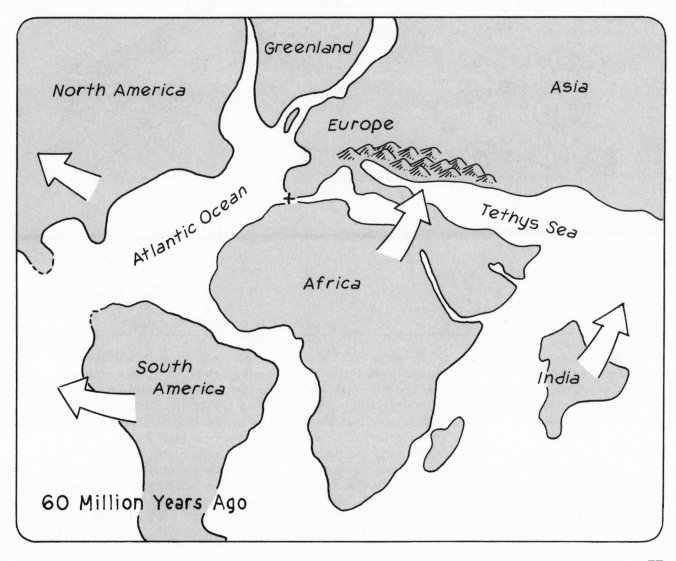

60 Million Years Ago

33 The Alps

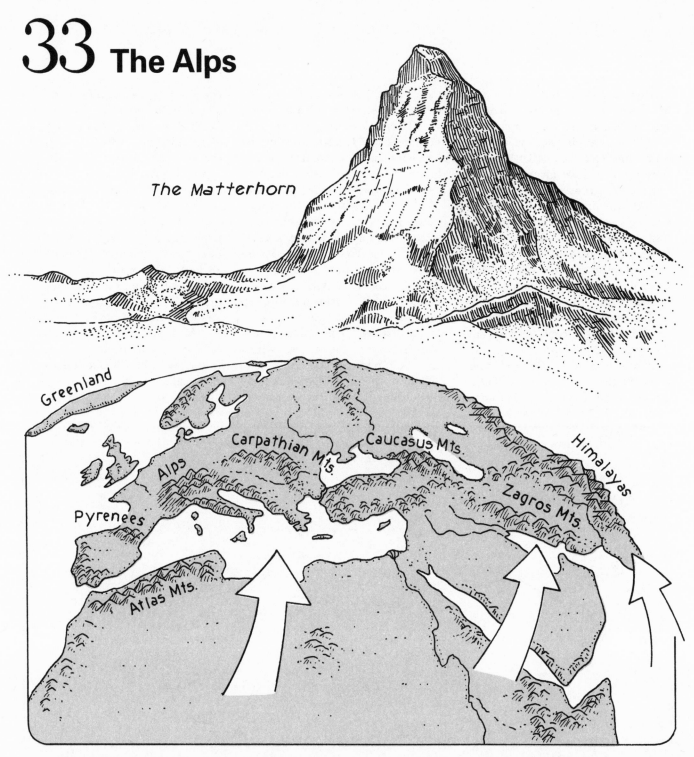

The Matterhorn

Greenland

Pyrenees

Alps

Carpathian Mts.

Caucasus Mts.

Himalayas

Zagros Mts.

Atlas Mts.

The Alps! The very name conjures up visions of lofty peaks and deep stream-cut valleys, of armies marching through snowbanked passes and shepherds tending flocks in high grassy meadows. No other mountain range on earth has been more extensively studied by naturalists, geologists and engineers. And no other mountain range has posed more perplexing puzzles of origin and structure.

It was in an Alpine valley that Louis Agassiz first caught sight of the ice ages of the past. And it was in the Alps that the fifteenth-century Swiss naturalist Felix Hemerli had a vision of the powerful forces that throw up the crust of the earth into towering Matterhorns. Hemerli found fossils in Alpine rocks, fossils of creatures that could only have lived in the sea. And he found them far from present day seas and thousands of feet above

sea level. It was not enough for Hemerli to follow his contemporaries and invoke the flood of Noah to explain these misplaced seashells. Instead, he surmised a stupendous crunching and folding up of the earth's crust, an explanation that in the fifteenth century must have seemed outrageously farfetched. But Hemerli was on the right track. Following his lead, geologists became increasingly aware that most of the great mountain ranges of the world have been thrown up from the beds of ancient seas. The cause of these gigantic upheavals was unknown.

In the eighteenth and early nineteenth centuries the puzzle began to deepen. Hans Conrad Escher and his son Arnold recognized that the Alpine rocks showed evidence of having been thrust northward into great overlapping folds, like "nappes" they said (from the French word for "tablecloth"). The term has remained in geology. The folding of the rocks piled older strata of limestone or metamorphosed sea-bed sediments on top of younger strata. Some of these nappes or overthrusts extend for tens of miles and involve beds of rock a mile thick.

The insight of the Eschers was strengthened when the first of the great Alpine tunnels, the twelve-mile-long Simplon Tunnel, was driven through the backbone of the Alps along the Swiss-Italian border in 1895-1905. The tunnel provided geologists with an "inside view" of the mountains, and confirmed that the overthrusts involved not only the surface formations but the very roots of the mountains themselves (see illustration below).

And yet, until recently the standard gospel of mountain building was that peaks were raised by essentially vertical movements of the earth's crust. How then came to be the overthrusts revealed by the Simplon Tunnel? Wegener's radical theory of large scale horizontal displacements would seem to have offered a more plausible explanation for the Alpine nappes, all of which are folded in the same direction as if pushed by some great force from the south. But acceptance of Wegener's ideas was delayed until the 1960s, when new and compelling evidence came not from the mountains but the sea floor. It was submarine topography and the record of the magnetic anomalies [25] that guaranteed posthumous fame for Wegener.

It now seems likely that the Alps (and the Pyrenees, Carpathian, Caucasus and Zagros Mountains) were pushed up when Africa, pivoting on Gibraltar, swung north, closing up the Tethys Sea and delivering Europe and western Asia a broadside blow. The floor of the vanished sea, thick with sediments, was driven below the margins of the northern continents or piled high on top of the continents in colossal folds and overthrusts. Italy, the Balkans and the Greek peninsula seem to be small fragments of Africa that had split away from that continent during the breakup of Pangaea, much as the Rockall Plateau broke away from northern Europe. These fragments were later driven like battering rams into the underside of Europe. The famous Matterhorn is a chunk of continental Italy that was pushed up onto Switzerland, with a base of Tethys sea floor rock caught in the fold. By this account, the Mediterranean Sea is the last remnant of the basins that divided the Italian and Balkan minicontinents from the African mainland, and it too seems destined to be consumed in the continuing crush of continents.

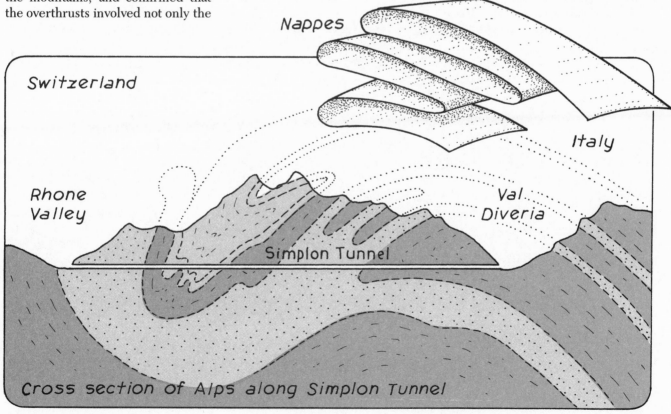

Cross section of Alps along Simplon Tunnel

34 Demise of the Dinosaurs

For over a hundred million years the dinosaurs ruled the crust of the earth. The first mammals appeared during the reign of the dinosaurs, and the first birds, but no other creature dared challenge the supremacy of the versatile reptiles. To insure their own survival, the dinosaurs relied on an astonishing array of adaptations. Some found safety in speed and agility, others in armor. Still other dinosaurs seem to have relied on sheer size to give themselves a competitive edge. A particularly large relative of the brachiosaurus may have stood fifty feet high and weighed a hundred tons—as much as a good-sized herd of elephants. Illustrated below are two successful dinosaurs of the late Cretaceous era, the heavily armored plant-eating Triceratops, and the undisputed king of the Cretaceous earth, the fierce meat-eating Tyrannosaurus.

Then, 63 million years ago the dinosaurs vanished, with surprising abruptness. Triceratops and Tyrannosaurus were not the only victims of this quick disappearing act. Many other species of plants and animals died out at the same time. But whatever the cause of these mass extinctions, the dinosaurs seem to have been the most susceptible victims.

Paleontologists have put forward a number of theories to explain the mass extinctions of the late Cretaceous era. Until recently, the most widely accepted theory attributed the extinctions to a cooling climate, perhaps caused by a rearrangement of continents and changing patterns of circulation in the oceans and atmosphere. This theory, like others which had been proposed to account for the demise of the dinosaurs, is not without problems of consistency and proof.

Another possible answer to the puzzle of the vanishing dinosaurs turned up recently in a layer of clay two centimeters thick separating two limestone formations in the hills near Gubbio, Italy. The limestone was laid down on the floor of the Tethys Sea and subsequently folded up to become the backbone of the Italian peninsula. The limestone below the clay layer contains marine fossils of the late Cretaceous period. There are no fossils in the clay. In the limestone above the clay there are fossils characteristic of the early Paleocene period. The age of the clay layer corresponds very well with the time of the dinosaur extinctions, and marks the geological boundary between the age of the dinosaurs and the age of the mammals.

In 1979, a group of scientists at the University of California at Berkeley revealed that the Gubbio clay contained an abnormally high level of the heavy element iridium, a concentration thirty times richer than in the adjacent strata. What could have been the source of this anomalous concentration?

Iridium is an uncommon element in rocks of the earth's crust, but it is more abundant in meteorites. This led the Berkeley group to suggest that 63 million years ago the earth was struck by a giant meteorite, perhaps as large as six miles in diameter. The environmental impact of such a collision would have been staggering. The shock wave alone could have killed large land-dwelling animals. If the meteorite struck in the sea, it might have raised a tidal wave several miles high which would have

Triceratops

washed over the continental margins. Dust or steam blasted into the stratosphere would have blocked sunlight for several years and drastically modified climate and the process of photosynthesis. This would have led to the collapse of food chains. It is easy to imagine that the exposed, highly specialized dinosaurs might have been easy victims of the catastrophe. The small resourceful mammals, scampering below the reptilian giants, may have been luckier.

The solar system is full of meteoric material. It has been estimated that a meteorite several miles in diameter will hit the earth every 100 million years, and so it would not be surprising to find evidence of a "hit" 63 million years ago. But the meteorite theory for the demise of the dinosaurs also has its problems. The data is not as complete or as consistent as one might like. No one has found a suitable candidate for the crater. And a careful study of the mineralogy of the clay layer suggests a volcanic origin for the fallout.

The dinosaurs may have been victims of volcanic convulsions and a changing climate. But anomalous iridium concentrations in 63-million-year-old strata have now been found worldwide, in places as diverse as New Zealand and Texas, all contemporary with the time of the mass extinctions. If the Berkeley group is right, Triceratops and Tyrannosaurus were dealt a fatal blow from outer space.

Tyrannosaurus

35 Last Days of Pompeii

The citizens of the Roman towns of Pompeii and Herculaneum never dreamed they had anything to fear from the bowl-shaped mountain that formed a pleasant backdrop to their markets, forums, amphitheatres, homes and shops. A few visitors to the mountain, including the Greek geographer Strabo, guessed that the peak had a volcanic origin, but the crater showed no sign of life and was thought to be extinct. The slopes of the mountain were covered with trees and vineyards. In 71 B.C. the green bowl within the rim served briefly as a stronghold for rebel slaves led by the gladiator Spartacus.

In A.D. 63 the mountain began to shudder, and earthquakes caused some damage in the nearby towns. The quakes continued year after year, but still no one guessed the mountain was a serious threat. On a sultry August afternoon in A.D. 79, the mountain came suddenly and decisively to life. The admiral-philosopher Pliny the Elder left his ship at anchorage near Pompeii to observe more closely the eruption. His young nephew recounted the fatal departure: "Ashes began to fall around his ships, thicker and hotter as they approached shore. Cinders and pumice and black fragments of rock cracked by heat fell around them. The sea suddenly shoaled, and the shores were obstructed by masses from the mountain."

Deep within the mountain's roots, in a chamber of molten rock, pressure had been long building. Now it explosively burst through the floor of the ancient crater, hurling huge quantities of ash and pumice into the air. The citizens of Pompeii were overwhelmed and buried by white

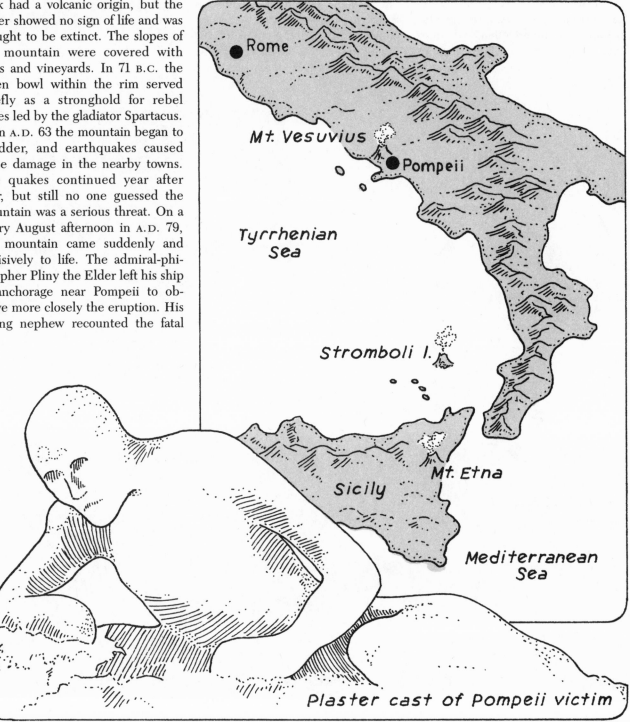

Plaster cast of Pompeii victim

82

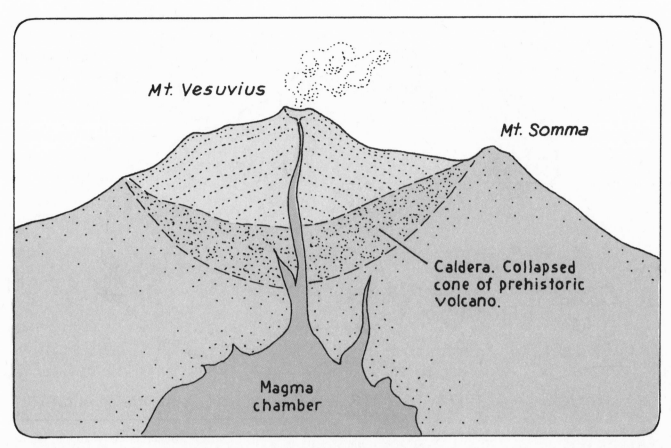

Mt. Vesuvius

Mt. Somma

Caldera. Collapsed cone of prehistoric volcano.

Magma chamber

hot ash mixed with steam that rained from the sky. Herculaneum was submerged in a sea of mud. Altogether on that terrible day as many as 16,000 people lost their lives, most by suffocation in the steaming ash.

The last agonies of the Pompeians have been remarkably preserved. Where the frightened citizens fell they were quickly covered by ash. Ultimately the bodies decomposed, leaving hollow cavities in the compacted ash that retained the form and posture of the victims. Today, archeologists excavating the buried city probe for these cavities and fill them with plaster of paris. When the matrix of ash is removed, a kind of statue remains, a macabre plaster replica of a Pompeian caught in terrified flight. The ash preserved details of musculature, facial expressions, even folds of clothing. The story told by these curious space fossils is of a populace which waited too long and realized too late the terrible danger they were in.

Since the time of the eruption of 79 A.D. the mountain has remained intermittently active, building up the present peak on the collapsed cone of an earlier prehistoric volcano known as Mt. Somma. A particularly violent eruption in 1631 took thousands of lives. The most recent awakening of the mountain occurred during the Allied invasion of Italy in 1944. The volcano will undoubtedly come to life again.

Vesuvius is the only active volcano on the European mainland, but it is one of a string of famous volcanic peaks near the toe of the Italian boot. Mt. Etna on Sicily and the volcano on the nearby island of Stromboli have been almost continuously active throughout recorded history.

The source of the heat that stokes the furnace of these mountains lies deep in the earth's crust. Along the line of the three active peaks two crustal plates are in collision. Africa is pushing the continental fragment

of Italy against Europe's underside, and driving the floor of what is left of an arm of the ancient Tethys Sea down beneath the northern continent. Heat generated along the subducting plate melts some of the overlying rock. The molten rock, called magma, is less dense than its surroundings. Where the magma finds or forces a path to the surface it pours out as lava or, if temporarily blocked, builds up the explosive pressure that can blast rock, ash and cinder into the sky.

In A.D. 79 at Pompeii the crust of the earth cracked open and the fiery devils of the colliding plates hurled up a mountain of pulverized earth. Falling on Pompeii, this blanket of ejecta snuffed out the life of the city and interred it for the ages. Excavated in our own time, the well-preserved ruins of Pompeii provide an extraordinary historical snapshot of a Roman town caught in one unhappy—and final—moment of its history.

36 Lost Atlantis

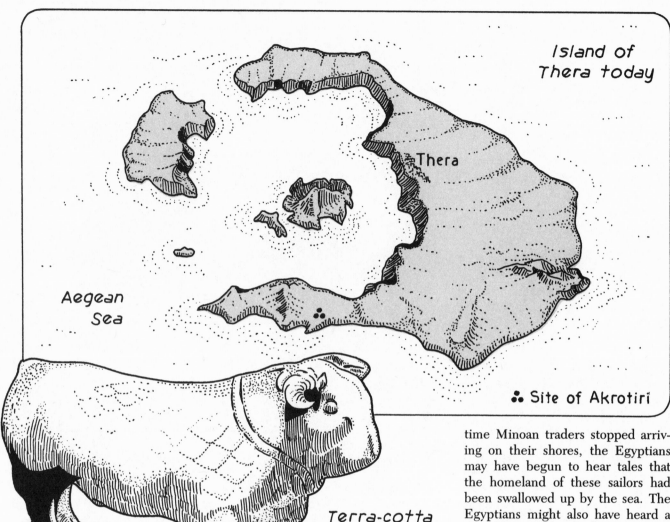

Island of Thera today

Thera

Aegean Sea

Site of Akrotiri

Terra-cotta bull from Thera

The legend of the lost continent of Atlantis is based on a tale told by Plato of a hostile nation that attacked Greece sometime in the distant past. These enemies of the Greeks met an awful fate, for, according to Plato, "There came terrible earthquakes and floods, and within the compass of a single day all of (the hostile) warriors were consumed by the earth, and the island of Atlantis disappeared beneath the waters of the sea."

Plato seems to have received the crude elements of his story of Atlantis by way of Egypt. What was the source of the Egyptian account? Many scholars are now confident that the Atlanteans of the Egyptian account were in truth traders from the island of Crete, to the north of Egypt across the Mediterranean Sea. The brilliant and successful Minoan civilization (whose sacred symbol, the bull, is illustrated here) flourished on Crete from 3000 B.C. until about 1450 B.C., when it collapsed with a suddenness and apparent violence that has long baffled archeologists. At about the same time Minoan traders stopped arriving on their shores, the Egyptians may have begun to hear tales that the homeland of these sailors had been swallowed up by the sea. The Egyptians might also have heard a great booming sound that rolled across the sea from the north. They were almost certainly showered by a mysterious ash that blew in on a north wind.

Geologists and archeologists have now pieced together a remarkable story about the end of Atlantis. It was not Crete which was consumed by the sea, but the island of Thera (sometimes called Santorini) seventy five miles north of Crete. Thera was an outpost of Minoan civilization, and from beneath many feet of ash archeologists have excavated an extensive Minoan city known as Akrotiri. The Therans were fishermen and farmers, with a thriving export trade to Crete. In return for the products of sea and olive groves, the Therans imported pottery and other goods and comforts from the mother island.

At first glance, the fate of Akrotiri seems similar to that of Pompeii, a prosperous city suddenly buried by ash. But few human remains have been found in the ruins of the Theran city. Unlike the Pompeians, the Therans had the good sense to evacuate their island before the catastrophe. But, as we shall see, their flight may have been in vain.

Thera is and was a volcanic island, one of the arc of islands thrown up above subducting sea floor plate where Africa and Europe are being squeezed together. About 1450 B.C. the long-sleeping island came to life, at first jolting the inhabitants with minor quakes. Then came a series of eruptions, followed at last by one of the most colossal volcanic explosions of all time. This final blast threw into the sky millions and millions of tons of ash and pulverized rock, which blanketed what was left of the island and was carried downwind across the sea toward Egypt. The cone of the volcano collapsed into the emptied magma chamber beneath the island, forming the deep round water-filled crater which is the distinguishing feature of the island today. Only a fraction of the original island remains.

The destruction of Thera may have brought about the end of Minoan civilization. A tidal wave hundreds of feet high must have battered the shores and harbors of Crete, dealing a devastating blow to the economic life of that seafaring nation (and, incidentally, perhaps swamping the citizens of Thera as they fled across the sea in their ships). The thick ashfall from the eruption would have destroyed Cretan crops and, for several years, rendered the soil unfertile. The weight of the ashfall may have been the mysterious blow from the sky, long wondered at by archeologists, that felled or damaged Minoan palaces, including those at the capital city of Knossos. Although the human toll on Crete may have been modest, the social and economic fabric of Minoan culture was delivered a deadly blow from which it never recovered.

The end of Minoan culture left a vacuum of power in the Mediterranean world which was quickly filled by the Mycenaean Greeks. And so began on the mainland the development of the culture from which European civilization so intimately springs. The course of history in the Mediterranean may have been significantly changed by the subduction of plates and the clash of continents.

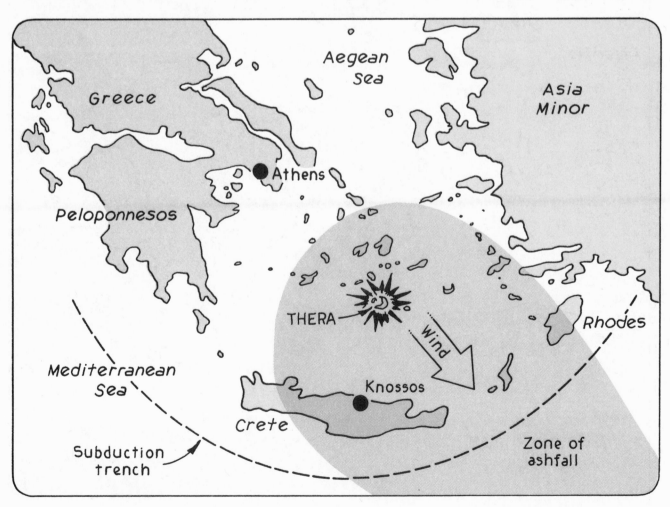

37 When the Mediterranean Was Dry

Every year 1000 cubic miles of water are evaporated from the Mediterranean Sea by the same warm sun that pale-skinned vacationers travel miles to seek. Only 100 cubic miles of this loss is replaced by rain falling onto the sea, or by rivers flowing into the basin. The deficit in the water budget is made up by a continuous and massive inflow of water from the Atlantic through the narrow Strait of Gibraltar. The Gibraltar connection is precarious—close it and the fate of the basin is sealed. A little arithmetic with the above numbers tells the story. Turn off the "faucet" at Gibraltar, and the "tub" of the Mediterranean will dry up in a thousand years.

In 1961 the American research vessel *Chain* explored the sea bed of the Mediterranean with a new kind of super-sonar. Powerful sound waves were bounced off the bottom sediments and recorded, providing a topographical profile of the sea floor. The survey revealed the presence of domelike structures that seemed remarkably like the salt domes of the Gulf Coast of the United States. Salt domes in shallow coastal waters are not unexpected. Domes are formed by the upwelling of rock salt from deeply buried beds of salt that were deposited by the evaporation of sea water in shallow coastal basins or lagoons. But salty evaporites are not to be expected on the bottom of a deep sea—unless that sea was once dry! The sound waves of the super-

sonar were able to penetrate the soft bottom sediments and revealed a continuous hard surface a few hundred feet below the sea bottom. This discovery suggested a bed of crusty evaporites such as might be seen at Bonneville Salt Flats in Utah [7]. The Utah salt flats are without question the residue of an evaporated inland sea.

A visit of the deep sea drilling vessel *Glomar Challenger* [48] to the Mediterranean Basin in 1970 confirmed the earlier suspicion. The hard reflecting layer buried in the sea floor sediments was indeed composed of salty evaporites. A study of samples brought up by the *Glomar Challenger* provided a remarkable history for the Mediterranean.

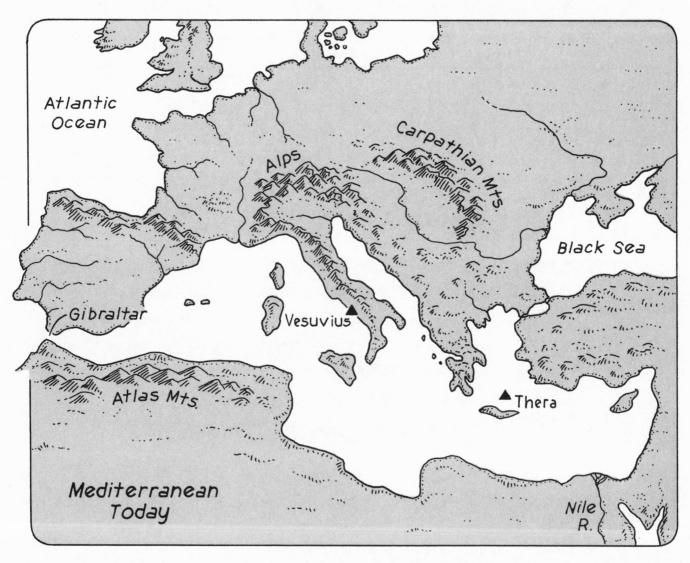

Atlantic Ocean

Alps

Carpathian Mts.

Black Sea

Gibraltar

Vesuvius

Atlas Mts.

Thera

Mediterranean Today

Nile R.

Sometime about 8 million years ago the continuing pressure of Africa against Europe pushed up highlands across the narrow Strait of Gibraltar and closed off the mouth of the Mediterranean. Without the Gibraltar "faucet" replenishing losses, the sun began to drink up the waters of the sea. Within a thousand years the Mediterranean was a barren, lifeless desert, many times larger, deeper and more inhospitible than Death Valley in California.

North of the dry Mediterranean basin, where the Black, Caspian and Aral Seas are today, there was at that time an inland sea known as Lac Mer, a cut-off arm of the old Tethys Sea. As rivers flowing into the dry Mediterranean basin eroded back into their headlands, channels were opened to Lac Mer. That body of water drained into the dead valley, and evaporated in brackish pools on the searing desert floor. Drainage from Lac Mer, and possibly occasional influxes across the Gibraltar sill, gave the Mediterranean a tenuous, halting life for millions of years. But finally the earth movements which raised the Carpathian Mountains changed drainage patterns and allowed the waters of Lac Mer to escape to the north (the Black, Caspian and Aral Seas are its lingering remnants). Again, and this time decisively, the Mediterranean went totally dry.

There have been suggestions that the drying up of the Mediterranean and the consequent deposition of huge quantities of salt may have diluted the saltiness of the world's oceans. Less salty water freezes more readily, allowing polar sea ice to extend and modifying world climate.

If any of the creatures who were humanity's immediate ancestors visited the brink of the two-mile-deep valley, the barren desert gulf must have seemed a more formidable barrier to travel than the former sea. Temperatures in the basin may have soared to a toasty 150 degrees F. Not even the heartiest forms of life could have survived long in the briny pools on the valley floor. Few more infelicitous environments have ever existed on the crust of the earth.

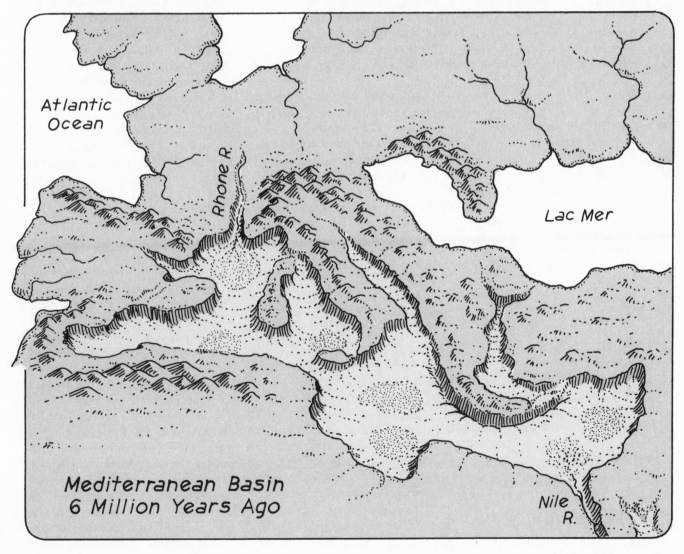

Atlantic Ocean

Rhone R.

Lac Mer

Mediterranean Basin
6 Million Years Ago

Nile R.

38 Grand Canyon of the Nile

When scientists from the research vessel *Glomar Challenger* announced their evidence for a dry Mediterranean they were greeted with almost universal skepticism. That one of the world's great seas, containing a million cubic miles of water, could evaporate completely like water in a pan seemed beyond credulity. But one response to the announcement was positive, and added further support for the theory. This came in the form of a letter from a Russian geologist who had been part of the team which planned the foundations for the high dam which the Soviet Union was helping the Egyptians build on the upper Nile at Aswan. The Russians had sunk a number of bore holes across the line of the proposed dam to determine the depth to granite bedrock. They were astonished to discover a deep gorge beneath the Nile Valley, a gorge now filled with sediments. The results of their survey are illustrated below. The narrow central gorge was cut into solid granite 700 feet below the present level of the Mediterranean Sea, and far below the present floor of the Nile Valley. Further north, in the Delta of the Nile, bore holes more than 1000 feet deep did not reach the bottom of the mighty buried chasm.

Even more puzzling, the deepest sediments that filled this canyon were of a marine origin and contained fossils of ocean-living organisms.

The discoveries which had been made by the *Glomar Challenger* provided a ready explanation for the buried "Grand Canyon of the Nile." When the Mediterranean went dry, the once sluggish Nile River was rejuvenated. Plunging vigorously down across the two-mile-high continental slope into the dessicated Mediterranean Basin, the river began to cut back into the continental platform. Eventually, the fast flowing river carved out a canyon that reached a thousand miles upstream, a canyon that rivaled the Grand Canyon of the Colorado in majesty. Where the river spilled out into the dry Mediterranean basin, it spread out and evaporated on the torrid desert floor, perhaps forming broad briny lakes during the season of floods.

When, some 5.5 million years ago, the Mediterranean refilled (see next pages), the Grand Canyon of the Nile became temporarily a long arm of the Mediterranean Sea, a watery dagger driven into the heart of Africa. It was at this time that saltwater fossils were deposited in the gorge beneath the site of the Aswan Dam. Now the river's work of erosion ended and deposition began. The valley began to silt up with sediments carried by the river from the interior. Finally the waters of the sea were pushed back into the Mediterranean Basin and the river built up a flood plain corresponding to the restored level of the sea.

Similar sediment-filled canyons underlie other streams flowing into the Mediterranean Basin. Near the end of the nineteenth century such a gorge was discovered beneath the valley of the Rhone River in France, cut into solid granite and filled with marine sediments. The existence of the buried Rhone gorge, and other buried canyons, constituted a geological mystery until the discovery of the dessication of the Mediterranean by the *Glomar Challenger*. That discovery was the key that suddenly unlocked an entire closet full of geological and paleobiological puzzles. For example, the drying up of the Mediterranean explained the radical change in climate that occurred in central Europe during the late Miocene, a time when tropical flora grew in Switzerland. And it explained abrupt disruptions in the fossil record of Mediterranean fauna which had long been recognized—but not explained—by biologists.

The idea of a dry Mediterranean basin had been suggested by other scientists prior to the voyage of the *Glomar Challenger*. Oil geologist Frank Barr had put forward essentially the same idea on the basis of his discovery of buried river channels in Libya. His paper was rejected by scientific journals as too outrageous. Italian paleobiologist Guiliano Ruggieri had argued for a dry Mediterranean on the basis of disruptions in the fossil record. His work, published in 1955, failed to receive serious attention for fifteen years. One must admire the intellectual courage of scientists such as these, who, like Alfred Wegener, saw the truth of a radical idea well before that idea's time had come.

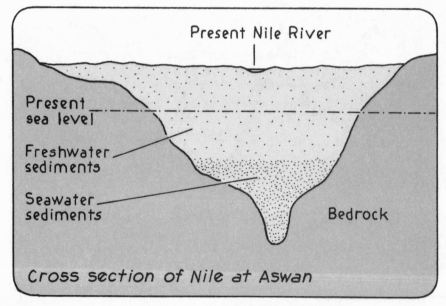

Cross section of Nile at Aswan

(labels: Present Nile River; Present sea level; Freshwater sediments; Seawater sediments; Bedrock)

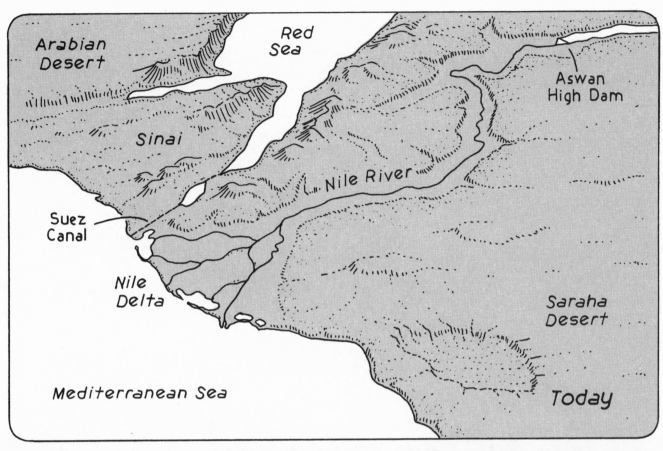

Arabian Desert

Red Sea

Aswan High Dam

Sinai

Nile River

Suez Canal

Nile Delta

Saraha Desert

Mediterranean Sea

Today

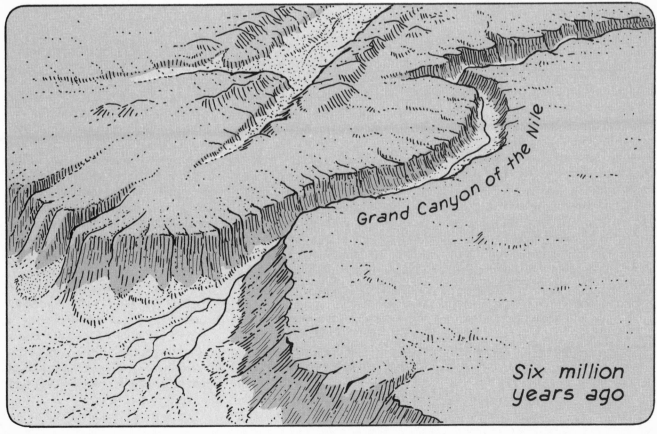

Grand Canyon of the Nile

Six million years ago

89

39 The Greatest Waterfall

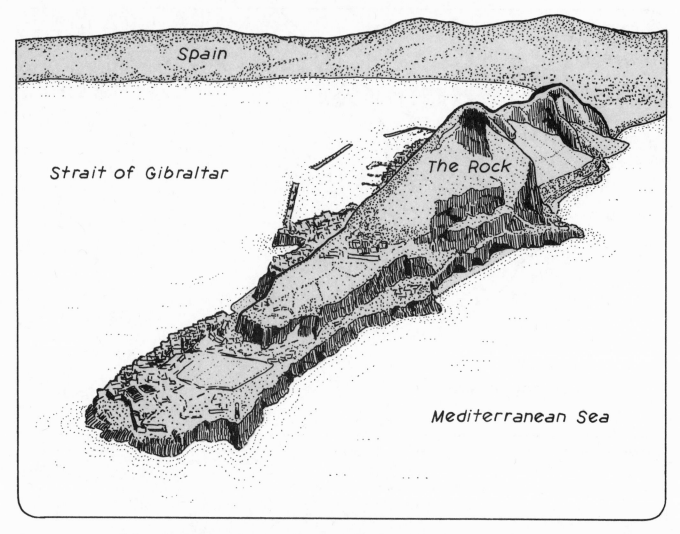

As an outpost of British Empire since 1838, the fortress rock of Gibraltar has withstood every assault by would-be conquerers. It has come to stand everywhere as a symbol for permanence. "As solid," the saying goes, "as the Rock of Gibraltar." But the rock is anything but solid. It is riddled with natural and artificial caves. The first Neanderthal skull was found in one of the natural chambers of the Rock. Other caves have yielded up the fossil bones of elephants and rhinoceroses.

Still, the Rock has endured as a geological as well as strategic fortress. It is a last bastion of the limestone wall that 6 million years ago held back the waters of the Atlantic and allowed the Mediterranean to dry up. The rock juts into the strait

between Europe and Africa like a great stone jetty. Its huge cave-riddled hump matches limestone formations on the opposite shore. The complete dike of which the rock was once a part may have been raised and broken several times during the late Miocene era, allowing the repeated drying out and filling up of the Mediterranean Basin. About 5.5 million years ago the barrier at Gibraltar was decisively breached and the basin has been filled with water ever since.

The existence of a precarious link between the continents across the mouth of the Mediterranean rested on a delicate balance between uplift and erosion. The pressure of the African plate against Europe folded up the floor of the Tethys Seaway to

create a limestone dam. The forces of erosion worked to tear down the rampart even as it was raised. The final collapse of the Gibraltar dam may have begun as a trickle, as the headwaters of streams flowing into the Mediterranean cut back across the isthmus toward the Atlantic. The trickle would have quickly grown into a thick stream, its eroding power enhanced by the steep gradient down to the floor of the dessicated sea. Limestone, as every spelunker knows, is easily eroded by water. The stream soon became a torrent, and finally a washout across the entire width of the strait. The resulting waterfall may have looked something like the very hypothetical drawing on this page.

Sea floor sediments brought up

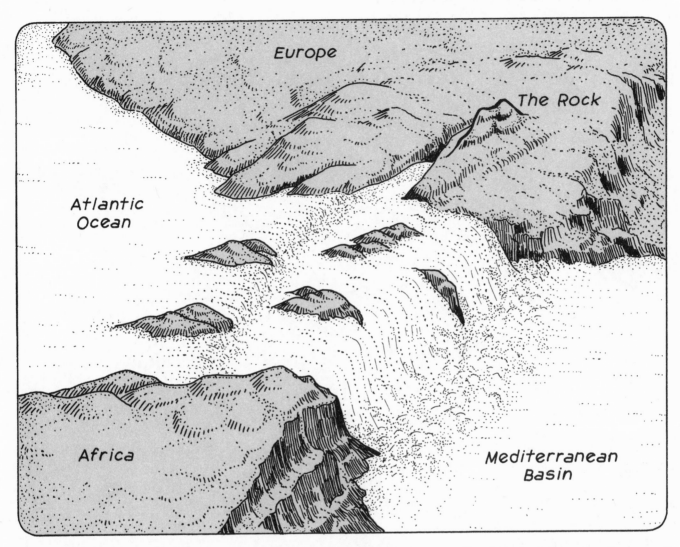

Europe

The Rock

Atlantic
Ocean

Africa

Mediterranean
Basin

from sites across the entire Mediterranean basin by the *Glomar Challenger* made it clear that the refilling of the sea was rapid. Simply to maintain losses to evaporation, the flow of water across the Gibraltar falls would have had to equal that of 100 Niagaras. To have provided enough water to support the fossil microorganisms found in the drilling samples, the falls would have had to pour the waters of 1000 Niagaras across the Gibraltar precipice. Even with this colossal rate of inflow, more than a century would have been required for the refilling of the entire Mediterranean Basin. In the process, the level of the world's oceans would have dropped thirty-five feet.

True human beings did not walk the crust of the earth until several million years ago. The Mediterranean refilled 5.5 million years ago. *Australopithecus*, an early ape-human, may have wandered to the edge of the abyss and witnessed the thundering of waters over the mile-high falls. One can only guess what dim stirrings of awe or imagination were sparked in that creature's expanding consciousness by the spectacle. The push and shove of continents continues, Africa still squeezes against Europe, and the floodgates at Gibraltar may close again. If a million years from now the cycle of evaporation and refilling is repeated, what sort of creatures will be standing on the diminished Rock of Gibraltar to watch the show?

Or will there be a show? Humans have the technological capability to

have breached the great isthmuses of the world—at Panama and Suez. Humans have become a significant factor in global geology and meteorology. Our technology is, wittingly or unwittingly, capable of modifying atmosphere and climate. We can tentatively predict and—if sufficiently determined—control volcanic eruptions and earthquakes. It is impossible to say what sort of creatures will dominate the earth millions of years from now—if life on earth has not destroyed itself in the meantime. But it is safe to say that the drying up of the Mediterranean will occur again only if those dominant creatures choose to allow it to happen.

40 A Green Sahara

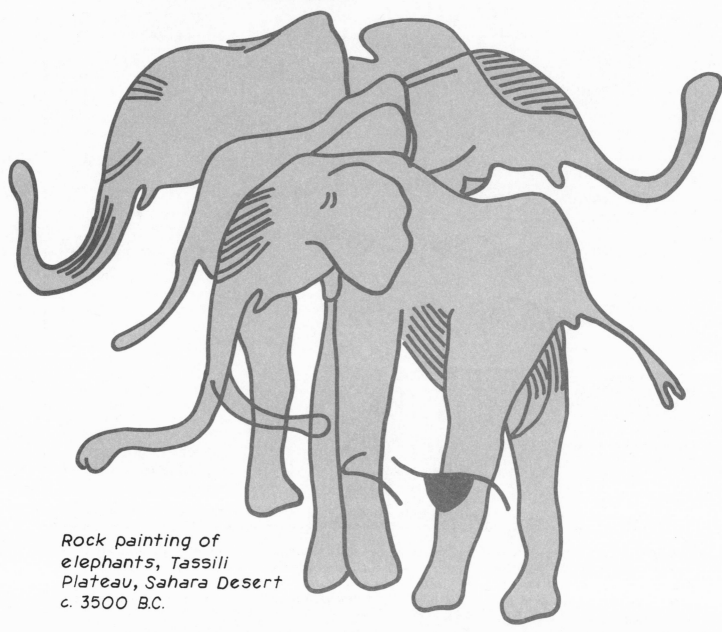

Rock painting of
elephants, Tassili
Plateau, Sahara Desert
c. 3500 B.C.

Sahara! Rolling dunes, sunbleached rocks. Camels and date palm oases. Salt caravans and watery mirages. Sahara! Arabs with flowing robes. The hidden city of Timbuktu. The French Foreign Legion. Sahara! Three thousand miles of blistering sand. The largest desert on the face of the earth.

Sahara! The name evokes heat and sand. On colored relief maps of the earth the Sahara appears as a great scorched blister. Across the entire breadth of North Africa, wind and sun have contrived a bleak, inhospitable wasteland. But it has not always been so. Climate is a fickle friend or foe. Not many thousands of years ago the Sahara was a fertile grassland, coursed by sparkling streams and teeming with wildlife.

Eight thousand years ago the final vestiges of the last great ice age still lingered in Europe [31]. The cooling of the earth that produced the continental glaciers, and the compounding effect of the glaciers themselves, moderated climate worldwide. Rains fell and water flowed in places where today there is only burning sand. The Sahara was green until about 2000 years ago, when changing climatic conditions allowed the Sahara to take on its present forbidding aspect.

During the long interlude when the Sahara was green, races of herdspeople thrived on the lush savannas, developing ever more advanced societies. They left a vivid record of life on the green Sahara in rock paintings of a wonderful deli-

92

cacy and beauty. Some of the best preserved of these works of art are found in the high central plateaus, most particularly at Tassili n'Ajjer. The paintings are the most complete record of stone age life to be found anywhere on the planet. Ironically, they have been preserved by the same dry air that brought about the decline of the culture they recorded.

The rock paintings of the high Saharan plateaus depict a grassland ranged by elephants and giraffes. There are images of a resourceful people who tended large herds of cattle, dug wells, and harvested wild grains. There are frescos depicting music and dance, religious ritual and battle. Today, these same caves of the central Sahara that were once the "Sistine Chapels" of a flourishing agricultural civilization are home to pit vipers and spiders.

In recent years geologists, paleobiologists and meteorologists have begun to piece together a dramatic new view of the terrestrial "weather machine." Clues have included the densities of fossil microorganisms in sea floor sediments, pollen grains buried deeply in the Antarctic and Greenland ice caps, growth rates of ancient coral reefs, and written records of the human species. We now recognize that the terrestrial weather system is like a finely tuned engine of astonishing complexity, subject in the smoothness of its running to minute variations of a hundred controlling elements. The changing orientation and distance of the earth with respect to the sun seems to be a crucial factor. Slight changes in the energy output of the sun, the drift of continents, rates of sea floor spreading, the tentative

opening and closing of crucial "valves" such as the Gibraltar strait or the Panama Isthmus, all exert their unsteadying influence. A change in average air temperature of only two degrees Centigrade is enough to trigger an ice age or dry up a Sahara. So delicate is the balance of forces that control the weather and so various are the contributing factors, that only the boldest meteorologist or outright oracle would dare to predict long term trends in the earth's climate.

But this much is clear from a study of the weather patterns of the past: we are living in one of the warmest periods of the past million years. The present Sahara Desert is possibly a product of what might be only a brief interlude, a kind of intermission, in a long, long age of ice.

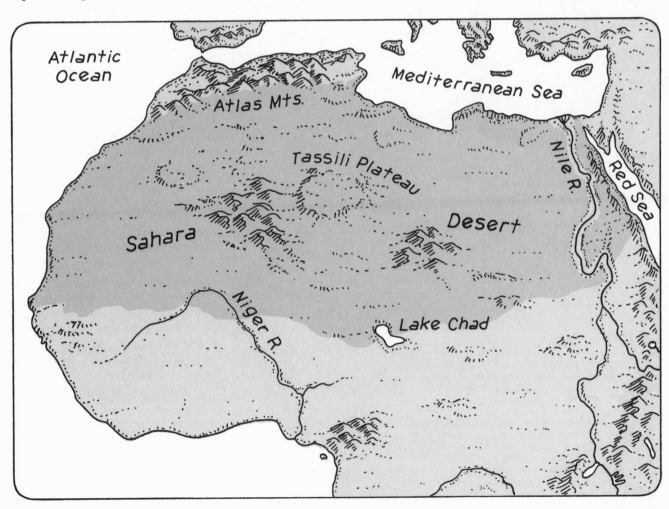

41 An Icebound Sahara

We have seen that 200 million years ago, when all of the continents were united in the super-continent of Pangaea, northwest Africa abutted eastern North America [19]. One could have walked from Boston to Morocco. But even further back in time, before the breaking up of Pangaea, what then? How long had that super-continent endured? Not long, apparently. The big crunch that pushed up the Appalachians came when Pangaea was assembled from earlier fragments. And in the Sahara Desert we get a glimpse into an even more distant past, a hint of the arrangement of continents 500 million years ago.

French geologists exploring for oil in the Sahara reported what appeared to be deep glacial scratches on outcrops of 500 million year old Ordovician sandstones. It seemed very unlikely that an ice cap once existed in what is today one of the hottest places on earth. Further investigation seemed called for. An international team of geologists set out in 1970 to follow up the French lead. The team criss-crossed the Sahara by plane and Land Rover, and found compelling evidence of heavy glaciation, always associated with Ordovician formations. Most striking of all were large areas of exposed sandstones deeply incised as if by moving ice. The orientation of the scratches, radiating from a center of apparent accumulation, suggested the presence of an ice cap such as presently

exists in Antarctica. The ice cap seems to have covered all of northwest Africa from Morocco to Lake Chad. Clearly, such an ice cap could not have formed in a tropical climate such as exists today in north Africa. Nor does it seem likely that a sufficiently cold climate could occur in tropical or equatorial latitudes. The implication was clear—500 million years ago north Africa was near one of the earth's poles. Drift subsequently carried the continent to its present location.

Other lines of research have confirmed the story told by the glacial scratches on African sandstones. For example, a study of the magnetism of volcanic rocks from the Ordovician period [24] points to a south magnetic pole, and therefore presumably the south geographic pole, somewhere in the vicinity of north Africa. But the story is far from complete and the reconstruction of continental positions in those distant times is a risky business. A tentative guess for the arrangement of continents on the Ordovician earth is shown at right. Africa is upside down with respect to its present position, and is situated at the earth's southern pole. It is part of a super-continent (Gondwanaland) that includes South America, Antarctica and India. Antarctica is tipped up near the equator, which suggests that the icebound continent might once have been a pleasant place to live [59]. North America lies on the equator. Indeed, rocks in eastern North America of the same age as the ice-scarred Sarahan sandstones contain fossils of marine organisms that could only have survived in a warm tropical climate. In upstate New York there are fossil coral reefs of the same age.

Five hundred million years is only a ninth of the age of the earth. Presumably the granitic continental masses of the earth's crust have arranged and rearranged themselves many times in the earth's history. As we peer into the geological reverse-

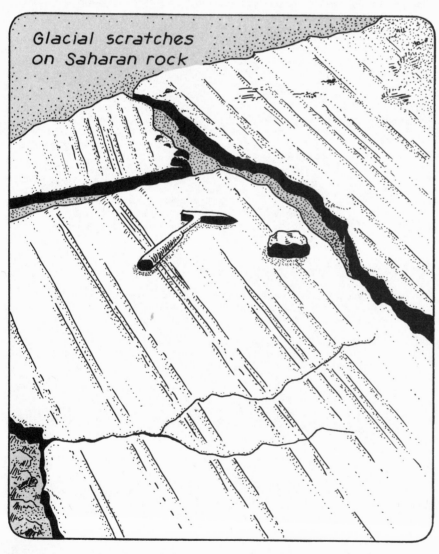

Glacial scratches on Saharan rock

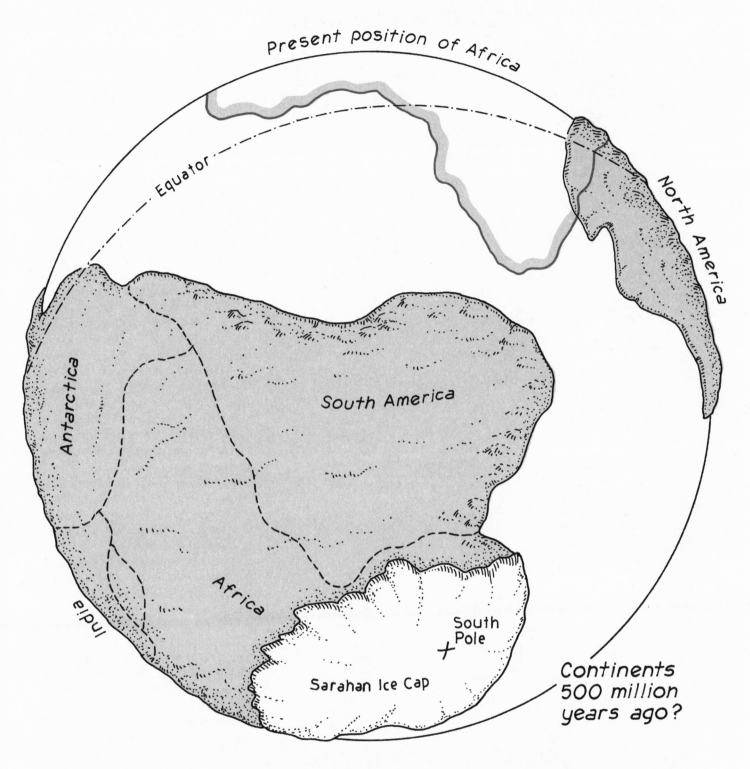

Present position of Africa

Equator

North America

Antarctica

South America

India

Africa

South Pole +

Sarahan Ice Cap

Continents 500 million years ago?

time crystal ball the picture becomes fuzzier and fuzzier the further back we look. The reconstruction on this page is about the earliest view of the earth's crust that can be sketched with confidence.

It would be nice to have a time-lapsed movie of our planet, made from the moon, say, that spanned the entire 4.6 billion years of earth history. It would show (we believe) continents barging about, colliding and breaking apart. It would show great mountain ranges being pushed up and worn down. And it would show ice sheets forming and melting, trop-

ical forests flourishing and falling into decay. In the absence of such a film, the best we can do is piece together the subtle clues offered by fossil reefs on mountain peaks and grooves in desert rocks. To tell the story of the past, the geologist must be a bit of a Sherlock Holmes.

42 The Birth of an Ocean

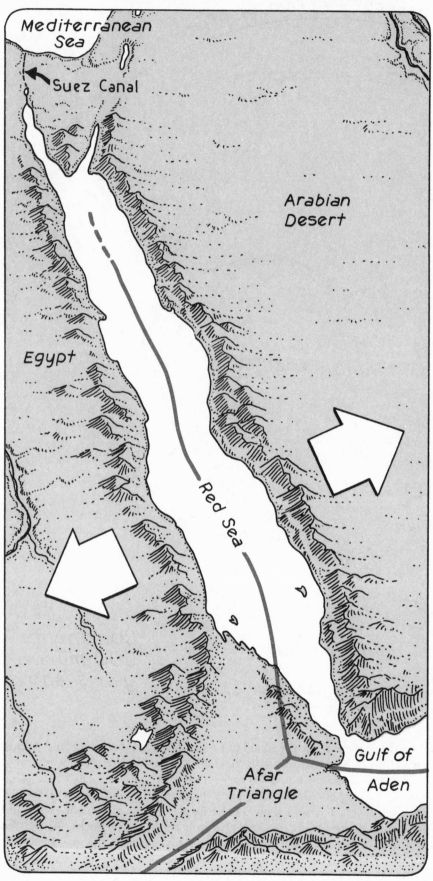

As astronaut Harrison Schmitt flew high above the Red Sea during the flight of Apollo 17, he radioed this message back to earth: "I didn't grow up with the idea of drifting continents and sea floor spreading. But I tell you, when you look at the way the pieces of the northeastern portion of the African continent seem to fit together separated by a narrow gulf, you could make a believer out of anybody."

One need not go into space to be impressed by the neat jigsaw-puzzle fit of the opposite coastlines of the Red Sea. A glance at the map will do. Only near the southern end of the sea does the parallelism of the coastlines break down, but even there the fit is remarkable if we include as part of the Red Sea basin the bleak depression known as the Afar Triangle. As we shall see (43), there are good reasons for doing so.

Geologists now regard the Red Sea as an extension of the rift system that cuts across the middle of the Indian Ocean basin (see map next page). Beginning about 20 million years ago, this rift opened up the Gulf of Aden. Then, turning a corner, the rift sliced up between Africa and Arabia, creating the Red Sea. Here, as elsewhere on the earth, forces within the earth's mantle are tearing apart the crust, splitting a continent into two parts.

In the Red Sea we may be witnessing a new ocean in the making. This may be what the South Atlantic Ocean looked like 150 million years ago when Pangaea split apart and a gulf widened between Africa and South America, opening from south to north like a geological zipper.

The forces that tear the earth's crust are believed to be related to convection currents in the earth's mantle. As the lighter continental rocks are fractured and pulled apart, fresh volcanic basalts well up from below to fill the rift. Because these basalts are denser than the continental granites, they do not rise as high as the continental platforms. Instead, they form the floor of a new and expanding ocean basin.

The floor of the Red Sea consists of salty evaporites and erosional debris washed down from the continental margins, deposited on dropped blocks of the continental platforms. No significant rivers flow into the sea, and the sea is connected to the Gulf of Aden by a shallow channel that is easily closed by a rise of the crust or a drop in sea level (as during an ice age). The presence of layers of salty evaporites suggests that the Red Sea dried up at various times between five and twenty million years ago. Drawings on pages 96 and 97 show the Red Sea in one of its dry intervals.

Down the long axis of the sea there is a mile deep trench floored with recent basaltic rocks which are typical of the rocks of the ocean crust. The very deepest parts of this rift have been found to contain pools of hot brine (water saturated with salt). The brines and the sediments below them are rich in iron, manganese, lead and other valuable minerals—tens of thousands of times richer than ordinary sea water. Ap-

parently hot sea water, percolating through sea floor sediments and fissures in the fresh volcanic rocks of the rift, have leached out soluble metals and concentrated them in the briny pools and sea floor fissures. These concentrations of metals may subsequently be incorporated into continents along subduction zones.

Many of the rich mineral deposits and ore veins of the world may have been formed along ancient mid-ocean rifts in a similar fashion. The Red Sea hot brines have great economic value, but their physical location at the bottom of the sea and the thorny legal question of ownership will make their exploitation difficult.

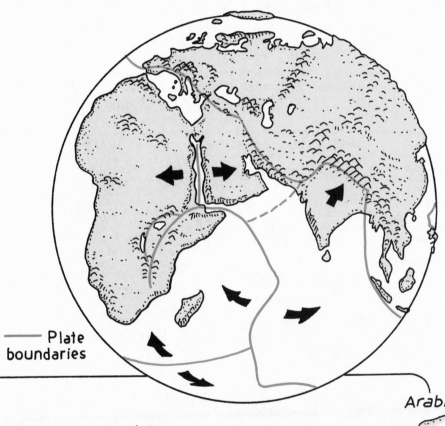

—— Plate boundaries

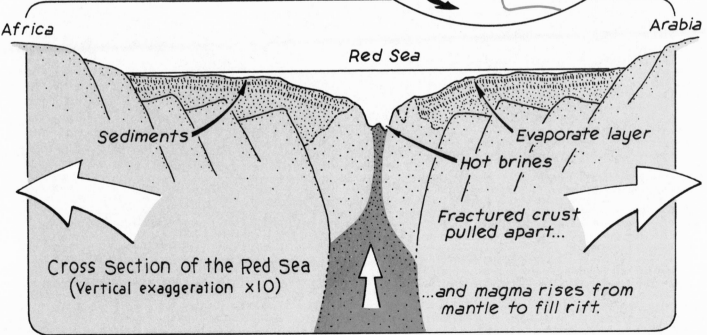

Africa

Arabia

Red Sea

Sediments

Evaporate layer

Hot brines

Fractured crust pulled apart...

Cross Section of the Red Sea (Vertical exaggeration ×10)

...and magma rises from mantle to fill rift.

43 The Afar Triangle

The jigsaw-puzzle fit of the opposite coastlines of the Red Sea is impressive evidence for the rifting of continents. Everywhere, that is, except at the extreme southern end of the sea. There the sea narrows to a strait only twenty miles wide, and the coastlines bulge toward each other in a way seemingly inconsistent with a geological jigsaw. Our first strong impression of continental rifting and sea floor spreading seems to go out the window. But wait! Let your eye wander away from the western shoreline and follow instead the line of steep escarpments that stand just inland from the Red Sea and Gulf of

Aden. At the southern end of the Red Sea these scarps turn inward from the coast. The two lines of cliffs meet 300 miles inland to enclose a wedge of low elevation land called the Afar Triangle. If we consider the Afar Triangle as an extension of the floors of the Red Sea and the Gulf of Aden, rather than as part of the African continent, then the fit of the opposite "coastlines" becomes remarkable, and the impression of rifting and spreading becomes strong indeed.

Geological surveys of the Afar Triangle have provided ample evidence to suggest that the area is in fact a

wedge of raised sea floor rather than true continent. An international multidisciplinary expedition of scientists carried out a detailed exploration of the area in the winter seasons of 1967-70. They found the terrain all but impassable, except by helicopter. The nomadic inhabitants of the area had a reputation for violence against visitors and were treated with respect and a wary eye. The results of the survey gave strong new impetus to newly emerging theories of ocean formation and continental drift.

The Afar Triangle is bounded on the west by the high Ethiopian Es-

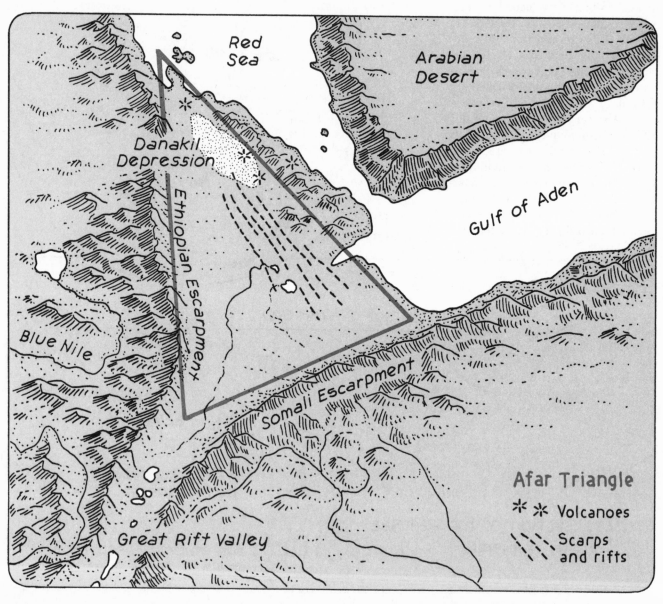

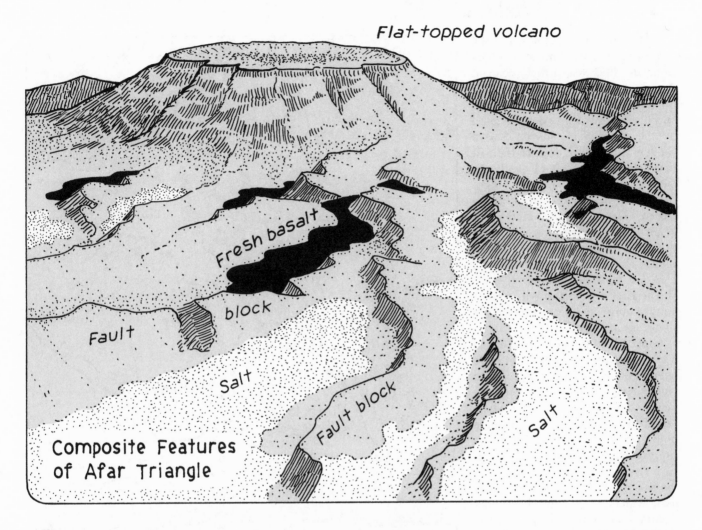

Flat-topped volcano

Composite Features
of Afar Triangle

Fresh basalt

block

Fault

Salt

Fault block

Salt

carpment, in some places standing more than 13,000 feet above the level of the sea. The cliffs of this escarpment rise higher and more abruptly from the valley floor than at any other place on earth. The southern side of the Triangle is defined by the mile high Somali Escarpment. At the northern vertex of the Triangle the broad and forbidding Danakil Depression drops 400 feet below sea level. The entire Triangle is so bleak, infertile and inhospitable to life that even today little is known of the geography of the area or the people who live there. The geological expeditions experienced temperatures as high as 134 degrees Fahrenheit in the summer shade and a sweltering 123 degrees in winter.

The Afar Triangle is the site of active and extinct volcanoes, hot springs and earthquakes. There are swarms of north-south tending faults, along which the crust has slipped to form cracks, volcanic dikes, and stairstep escarpments. All of these active tectonic features have trends running parallel to the axis of Red Sea. The rocks of these fault blocks are typical of the basalts of the ocean crust along sea floor ridges. There are flows of fresh basalt which have only recently welled up along fault lines. And there are flat-topped volcanoes and low cones of glassy volcanic materials that could only have been formed beneath the sea.

If the geologists are right, the Afar Triangle is a slice of freshly baked sea floor that has been temporarily pushed up above the waters at a point where three great rifts intersect. It may be that a "hot spot" of rising magma underlies the Triangle, an upwelling plume of mantle material pushing on the crust from below like a hot liquid finger. It may also be that the hot spot is itself partly responsible for the three-branched rift in the earth's crust.

There are only a few places on the crust of the earth where one can walk on "ocean floor." Iceland is one of them [23]. The Afar Triangle is another. Until quite recently the land of the Triangle lay beneath the waters of the Red Sea and Gulf of Aden. In many places there are thick deposits of salty evaporites only a few million years old. The stone artifacts of early humans have been found here encrusted with seashells. If the Red Sea continues to widen, the Triangle may again sink and become true sea floor. In the meantime, the Afar Triangle offers geologists a rare chance to study on dry land an ocean floor in the making.

44 The Great Rift

The Afar Triangle is at the nexus of a three-armed rift. One arm opens along the Red Sea, a new ocean in the making. North of the Red Sea this arm of the rift slices into Asia, forming the low depression occupied by the Dead Sea and the Sea of Galilee. A second arm of the rift reaches out along the Gulf of Aden to join up with the Carlsberg Ridge, the spreading axis of the Indian Ocean. The third arm of the rift begins at the southwestern vertex of the Afar Triangle and cuts down across East Africa, threatening to tear that continent asunder.

The East African Rift is a complex array of tensional faults, rather than a single clean fracture. It may be an example of the initial stage of continental rupture, such as might have been seen along the east coast of North America when the Atlantic began to open up 200 million years ago. The crust of the earth is being pulled apart in East Africa, but the rupture has been slow and indecisive. During the last 10 million years the valleys of the East African Rift have widened by only about one-twentieth of an inch per year, rather than the inch or so per year that is typical of mid-ocean ridges.

The rift array in East Africa seems to lie on top of a great blister or arch in the earth's crust. Much of the area has been lifted by thousands of feet. The uplift seems to be caused by an underlying mass of molten or nearly molten mantle material, which, because it is less dense than the surrounding rock, has expanded and lifted the crust. This rich magma source lies just beneath the crust, and is the source of the heat that warms the hot springs that bubble up along the rift valleys. It also stokes the volcanoes that stand nearby. Mounts Kenya and Kilimanjaro are among the largest volcanic peaks in the world. Kilimanjaro soars five times higher than Vesuvius.

Along the principal faults in the East African array, linear blocks of

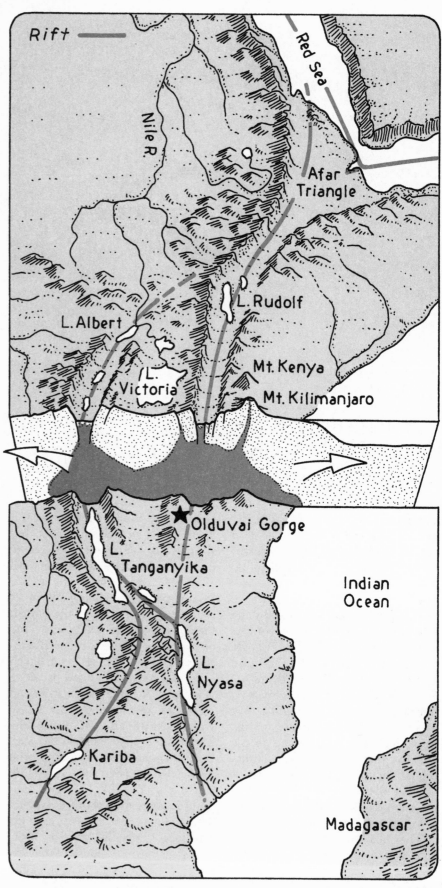

crust have dropped to form deep steep-walled, lake-filled valleys that give the region its spectacular scenery. The largest of these lakes, Lake Tanganyika, is the second deepest lake in the world, after Lake Baikal in Siberia.

There is no way to predict with certainty the future development of the East Africa Rift. A reasonable guess for how the region will look in 50 million years is shown on the map at left. Africa has continued to pivot northward about Gibraltar. The Mediterranean Sea has narrowed, crushing Italy more deeply into Europe. Perhaps the Mediterranean is closed off again at Gibraltar, but continued widening of the Red Sea has opened a new inlet at Suez to keep that shrinking basin full. The Arabian peninsula has pushed even more forcefully against Asia [46], squeezing the Persian Gulf out of existence. India has crunched onward toward China, maintaining the high range of mountains on its northern boundary [47]. And in East Africa continued rifting has opened a new arm of the Indian Ocean, not unlike the Red Sea and Gulf of Aden of today. East Africa has been set adrift as an island continent.

It is interesting to speculate about how life will adapt to these changing geological conditions. The British biologist Dougal Dixon has provided a fascinating account of life on earth 50 million years from now in his book *After Man*. The hypothetical cleft-back antelope illustrated below is adapted from Dixon's book, shown here with the symbiotic companions that nest in its cleft. Evolution of the antelope's spine from a mere perch for birds to a safe nesting bower has cemented a relationship that is useful to both bird and antelope. The birds eat insects kicked up by the hooves of the antelope, and ticks and mites from its hide. The antelope is rid of insect parasites and gains an early warning system—the screech of birds—in case of danger. According to Dixon's imaginative speculations, the cleft-back antelope is a rare survivor from the "Age of Man," only moderately evolved from the kind of ungulates that once roamed the East African grasslands in great herds. The cleft-back antelope's ecological niche was preserved by the rifting of East Africa from the mainland, providing the creature a kind of sanctuary. Elsewhere on the African continent, according to Dixon's delightful futuristic fantasy, the ungulates have been replaced by deer-sized, fast-running descendents of the rabbits!

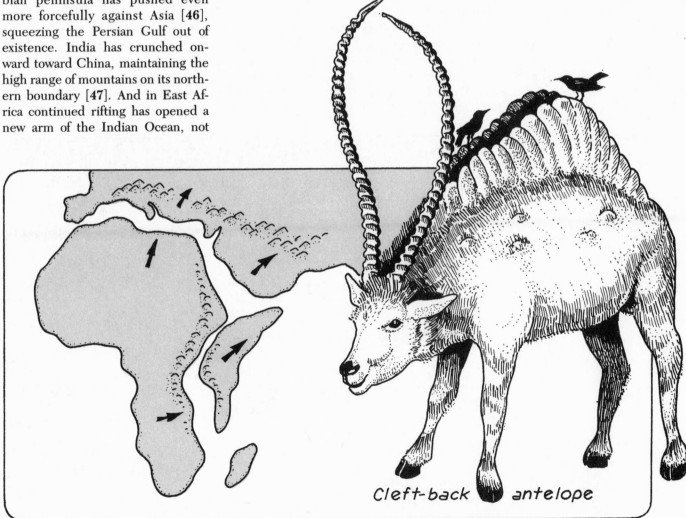

Cleft-back antelope

45 Fossil Footprints

No place has a better claim to be called the cradle of humanity than the rift valleys of East Africa, an array of steep-walled breaks in the earth's crust that slices down across the continent from the Afar Triangle in the north to the Zambezi River in the south. No other place has yielded up from the soil so many fossil bones and stone artifacts of our remote ancestors. The greatest number of these clues to human ancestry have been turned up in or near the Olduvai Gorge in Tanzania, 100 miles southeast of Lake Victoria (see map page 102). The story of the Olduvai Gorge is closely linked with Louis and Mary Leakey, a husband and wife team of anthropologists who spent decades of dedicated effort painstakingly sifting ash and sand. Their sons later joined them in this heroic exploration of the human past.

A good part of what we know about human evolution has emerged from the crust of the earth under the patient direction of the Leakeys. We know that humanlike creatures lived in East Africa millions of years ago. And it now seems certain that these creatures witnessed many of the geological events, volcanic eruptions and earthquakes, that accompanied the making of the rift valleys. A bit of the history of early human habitation of the rift valleys has been reconstructed from fragments of fossilized bone and stone tools unearthed in sedimentary strata that can be dated geologically.

One of the most exciting discoveries made by the Leakey team (unfortunately, after the death of Louis) came in a completely unexpected form. In 1976, beautifully preserved fossil footprints of hominids (human ancestors) were excavated in ash beds at a place called Laetoli, not far southeast of Olduvai Gorge. The ash could be dated rather precisely by radiometric methods to about 3.5 million years before the present. The distribution of the ash suggested that its source was the volcanic peak not far to the east known as Sadiman. The eruption of Sadiman that spread this layer of ash across the grassy plain predated the faulting that produced the nearby escarpments and created the basin now filled by Lake Eyasi.

The Laetoli footprints were preserved across the millions of years by an unlikely set of circumstances. Two hominids, perhaps an early form of *australopithecus*, walked one behind the other, or possibly at different times, across damp ash from a recent eruption of Sadiman. What they thought of the booming, fire-belching mountain can only be surmised. The ash hardened. Then a fresh ashfall covered the footprints before they could be eroded. Continuing deposition over the thousands of years buried the footprints under as much as sixty feet of sediments. There were episodes of faulting and uplift. At last, erosion exposed the 3.5 million year old ash layer and the footprints. Such a set of circumstances could only have occurred in a geologically active environment.

The footprints are shown in the composite drawing below. The tracks of a hare and a guinea fowl are also

Fossil footprints at Laetoli

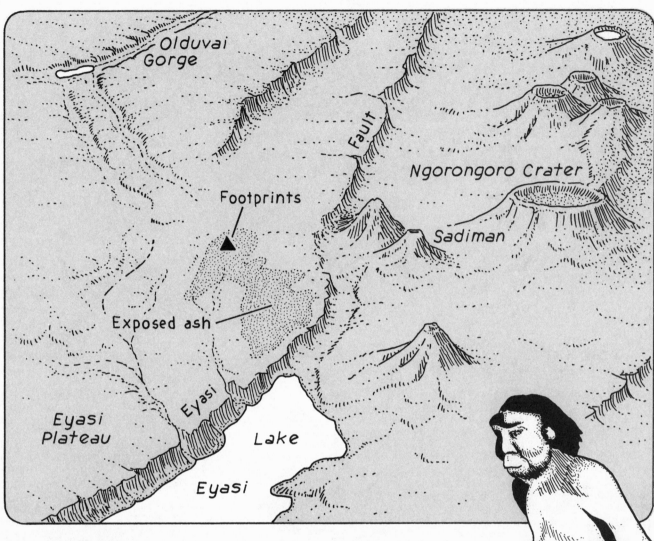

shown. The ash beds at Laetoli also revealed tracks of elephants, giraffes, rhinoceroses, antelopes and ostriches, a population not unlike what might be found in the same region today. Tens of thousands of tracks have been unearthed. The hominid footprints look remarkably fresh. And remarkably modern. They show a rounded arch, rounded heel, pronounced ball and forward pointing big toe, all features necessary for walking erect.

There is something especially intriguing about the Laetoli footprints. They are not so substantial as a fossil skull, or a stone chopper, or a piece of bone that might have been used as a tool or weapon. Yet in their own ephemeral way, they record flesh, life, activity, even (possibly) compan-

ionship. From the nature of the prints and the gait, it is possible to deduce that the prints were made by a male and a female. The male was about four-and-a-half-feet high, the female about four feet high. In the same ash beds with the footprints were found fossil parts of the skeletons of two adults and a child. It is tempting to suppose that these were the creatures that made the prints.

No stone tools have yet been found in these 3.5 million year old beds of ash. It seems that in the Laetoli footprints we have caught human evolution at that special moment between the time walking erect freed the hands for creative activities—and the discovery of what it was possible to do with them.

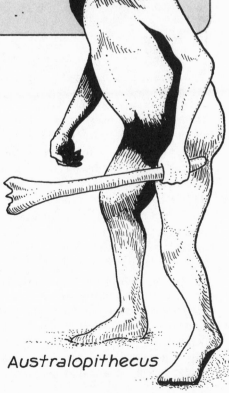

Australopithecus

103

46 Of Oil and Sand

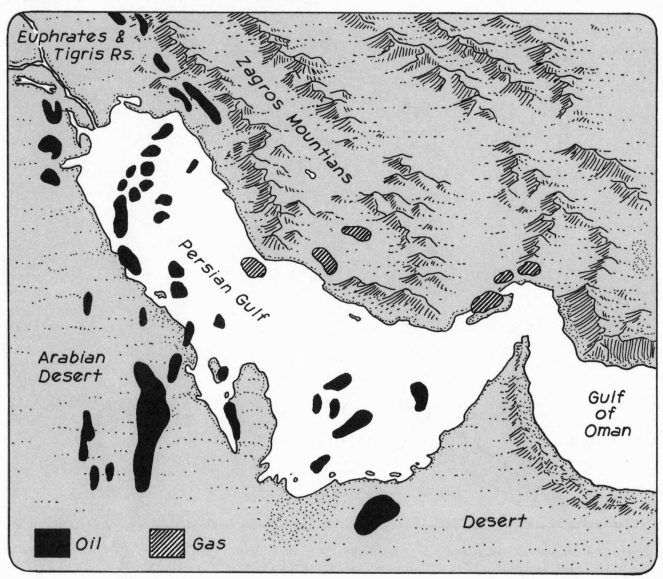

Euphrates & Tigris Rs.

Zagros Mountians

Persian Gulf

Arabian Desert

Gulf of Oman

Desert

■ Oil   ▨ Gas

The earth distributes its gifts in strange ways. Consider the patch of crust within the borders of the tiny desert nation of Kuwait, near the mouth of the Euphrates and Tigris Rivers at the top of the Persian Gulf. It is a barren place, a sandy wasteland bleached by the fierce sun. Drinking water must be distilled from the salt water of the Gulf and food must be imported. And yet Kuwait is today one of the wealthiest nations on earth. Its citizens enjoy free schools and free medical care and pay no taxes. The source of Kuwait's wealth, like that of the other desert nations of the Persian Gulf, lies beneath the sand. The

huge oil reserves of the Gulf are a hidden and unexpected treasure.

The oil resources of the Persian Gulf are the product of a complex set of geological circumstances. For oil to accumulate, several conditions are necessary. First, there must be a *source rock* for the petroleum. This is often a series of strata rich in the remains of single-celled organisms that thrived in ancient seas. In the case of the Persian Gulf, these source rocks consist of an almost uninterrupted sequence of organic-rich sediments interbedded with limestones and salty evaporates. These sediments date from a time several hundred million years ago

when the region lay at the margin of the Tethys seaway which divided the northern supercontinent of Laurasia from the southern supercontinent of Gonwandaland [32]. A second condition necessary for the production of petroleum is a porous *reservoir rock*, usually sandstone, fossil reefs or limestone, to provide passage and storage for the oil that is formed from the organic materials. Third, there must be an impermeable *cap rock* to contain the light upward-rising hydrocarbons and keep them from escaping to the surface even as they form. Fourth and last, there must be some kind of downward-concave *structure* to trap and hold the oil as in a kind of container. In the Persian Gulf, the trapping structures were created when the oil-producing strata were folded into domes and arches by plate collision.

The Arabian plate is one of many small lithospheric plates that are part of the "cracked eggshell" of the earth. Within a span of less than 1000 miles it displays most features of plate tectonics. The west side of the plate is the spreading center on the axis of the Red Sea. Here magma rises from the mantle to form new sea floor as the African and Arabian plates move apart. The eastern edge of the Arabian plate is a convergent boundary. Along the Persian Gulf the Arabian plate is diving beneath the Eurasian plate, crumpling the margin of the Eurasian plate and raising the Zagros Mountains. Here

the bed of the old Tethys seaway, with its thick organic-rich deposits, is being squeezed into folds and domes that contain and hold reservoirs of oil and gas. The squeezing pressure of the plates may help supply the conditions of heat and compression necessary to bring about the chemical reactions that change organic deposits into hydrocarbons.

The first oil well in the region of the Persian Gulf was sunk in 500 B.C. during the reign of Darius the Great, but massive exploitation of the petroleum resources did not begin until the 1930s. By 1975, the Gulf was producing over a third of the world's crude oil. Modern civilization has developed an insatiable appetite for energy. The oil and gas reserves of the Gulf have given that area an economic power and strate-

gic importance which would have seemed farfetched only a few decades ago. The resources of the Gulf have been 200 million years in the making. Yet it is possible that the energy-hungry species *homo sapiens* may deplete these resources in less than a century.

It also remains to be seen whether the massive exploitation of petroleum reserves in the Persian Gulf is compatible with the long-standing fisheries of the region. The shallow waters of the Gulf allow a high level of plankton production, which in turn supports a rich population of sardines, anchovies, mackerels and barracudas. It is an ecological balance which could be easily disrupted by the careless application of technology.

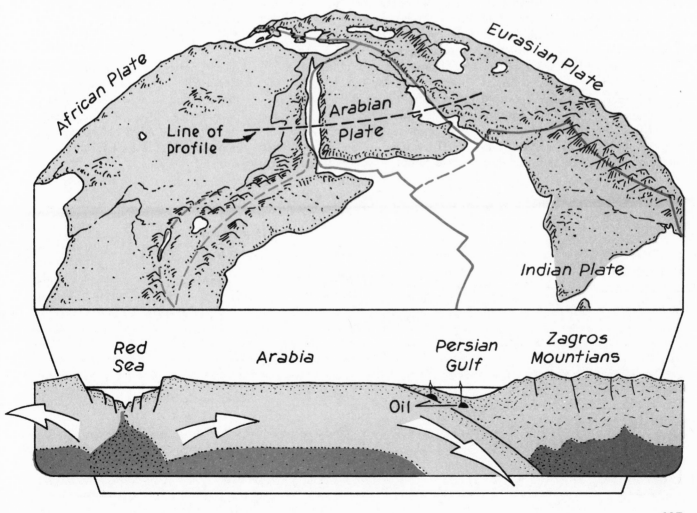

47 India on the Run

No part of the earth's surface has a more spectacular geography than the region north of India. Here are the highest mountains on earth, the Himalayas, rising abruptly from the Ganges plain to the awesome five-and-a-half-mile-high summit of Everest. Behind these lofty sentinels is the Tibetian plateau, the remote and inaccessible land of legendary Shangri-La. The average elevation of this broad upland is greater than Mt. Whitney in California, the highest point in the contiguous United States. Still further north is the inhospitable Gobi Desert, ringed on three sides by towering ranges. Beyond these, Lake Baikal, the deepest lake on the planet, fills a huge crack in the earth's crust. The entire region is frequently rocked by powerful earthquakes.

Geologists have long pondered what force could have so dramatically crumpled and thrust up this broad region of the earth's surface. The answer has finally emerged from the theory of plate tectonics, and in particular from the study of fossil magnetism [24,25]. Beginning about 40 million years ago and continuing today, the subcontinent of India has been involved in a colossal collision with the underside of Asia, throwing the margins of both crustal masses high into the air!

The story begins with the fragmentation of the supercontinent of Gondwanaland some 150 million years ago [32]. At that time India was tucked near the crustal slabs that would become the continents of Africa and Antarctica. First, a rift opened between India-Antarctica and Africa. Then, about 100 million years ago, India parted company with Antarctica and began drifting north. A new ocean basin, the Indian Ocean, opened south of the drifting continent, and the floor of an older sea which separated northern and southern continents was pushed down into the mantle beneath the southern perimeter of Asia.

Finally, beginning about 40 million years ago, continental India and Asia came into contact. Remember that continental rocks are less dense than those of the ocean floor and that the continent of India was riding like a passenger on a moving plate. The continent was too thick and too buoyant to follow the diving sea floor down into the mantle. As the continents came into contact, their margins were crumpled and elevated.

The climax of the great crunch came about 10 million years ago.

Since that time the growth of the Himalayas has continued to outpace erosion. Some geologists believe that what we have now is like a "double continent," one continent having partially underridden another. The breaks and gashes in the fractured crust are the source of the area's frequent and severe earthquakes.

Fossil magnetism has recorded the northward drift of India. The angle of inclination of the earth's magnetic field with respect to the surface of the earth is a function of latitude: for example, the field is nearly horizontal at the equator, points directly into the ground at the north magnetic pole, and so on (see illustration page 60). The drift of India across the equator from southern latitudes to its present northerly position is preserved for geologists in the fossil magnetism of Indian lavas.

Two continents have been sutured together along the Himalayas to form a new and larger continent. Clearly the northern movement of the India plate cannot proceed much further. Eventually, erosion will outpace uplift and the Himalayas will be "cut down to size," like the Appalachians, the Urals and other old ranges that mark the lines of earlier continental collisions.

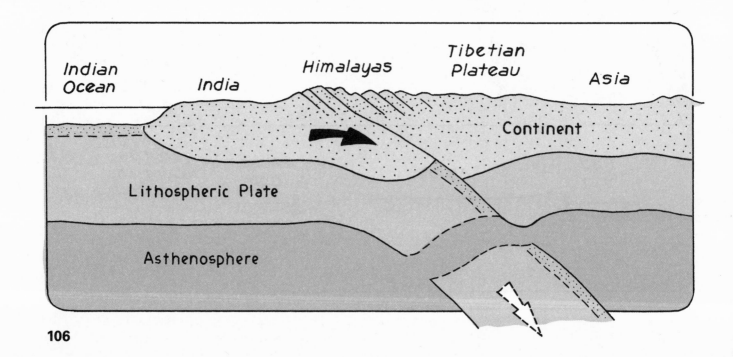

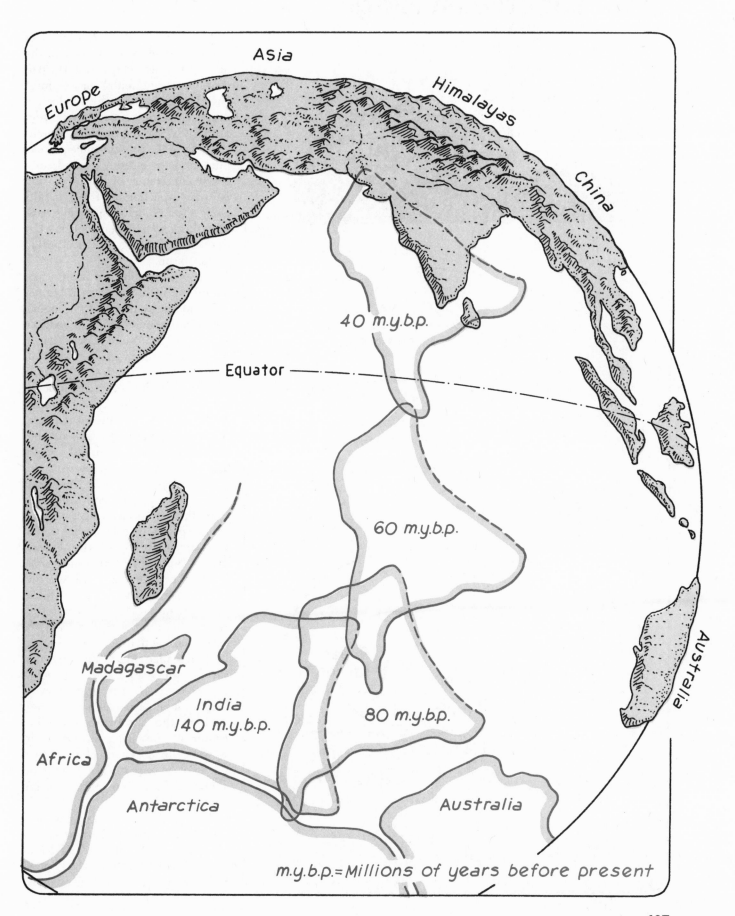

Asia

Europe

Himalayas

China

40 m.y.b.p.

Equator

60 m.y.b.p.

Madagascar

80 m.y.b.p.

India
140 m.y.b.p.

Africa

Antarctica

Australia

Australia

m.y.b.p.=Millions of years before present

48 Ninetyeast Ridge

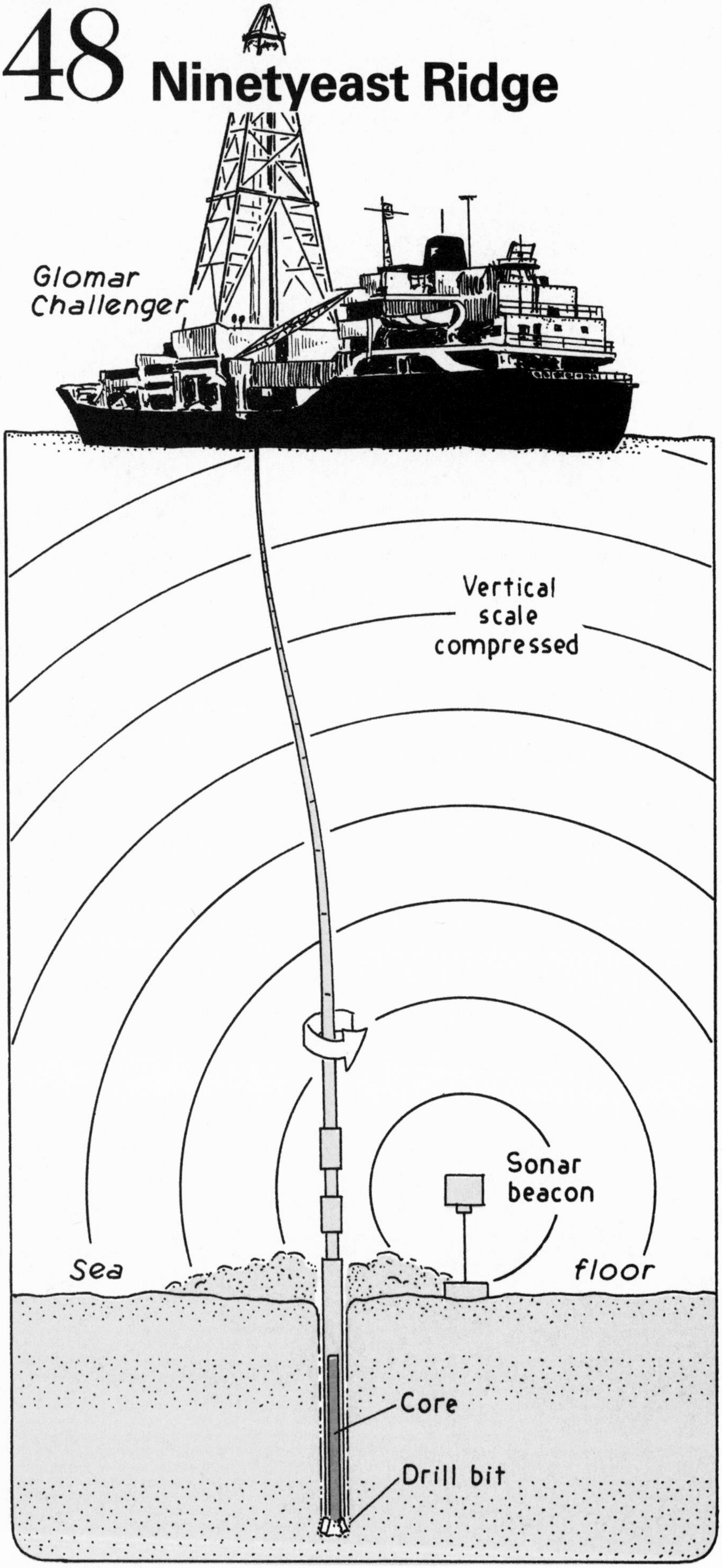

Glomar Challenger

Vertical scale compressed

Sonar beacon

Sea

floor

Core

Drill bit

If India has drifted thousands of miles to collide with Asia, then the floor of the sea which once divided those two continental masses must have been consumed by subduction into the mantle beneath the present Himalayas. By the same argument, the floor of the present Indian Ocean must be new, entirely created since the disintegration of Gondwanaland by the extrusion of fresh basalts along the mid-ocean spreading axis. That this is so has been confirmed by deep-sea drilling in the Indian Ocean Basin.

Some of the most dramatic topographical features on earth are forever hidden by the sea from human view. One splendid piece of invisible scenery is the Ninetyeast Ridge, a mile high submarine mountain range that runs for 2500 miles along a north-south line on the floor of the Indian Ocean. No other mountain range on earth is quite so perfectly linear. The range takes its name from the fact that it lies almost exactly along the line of longitude 90 degrees east of Greenwich. Ninetyeast Ridge is of a very different character from the mid-ocean ridges that lie along spreading axes, such as the Carlsberg Ridge of the Indian Ocean. Ninetyeast Ridge runs parallel to the direction of plate drift, rather than transverse to it. Unlike the spreading ridges, Ninetyeast Ridge shows no sign of volcanic or earthquake activity. Nor is it broken by a mid-ridge rift or transverse fractures. Ninetyeast Ridge was probably created by the extrusion of basalts at a particularly active hot spot on the mid-ocean spreading center and has been carried northward by the movement of the plate of which it is a part, like toothpaste squeezed from a tube onto a moving conveyor belt.

In 1972 the deep-sea drilling vessel *Glomar Challenger* sank five bore-holes along the crest of Ninetyeast Ridge. The ship was equipped to bring up thirty-foot long cylindri-

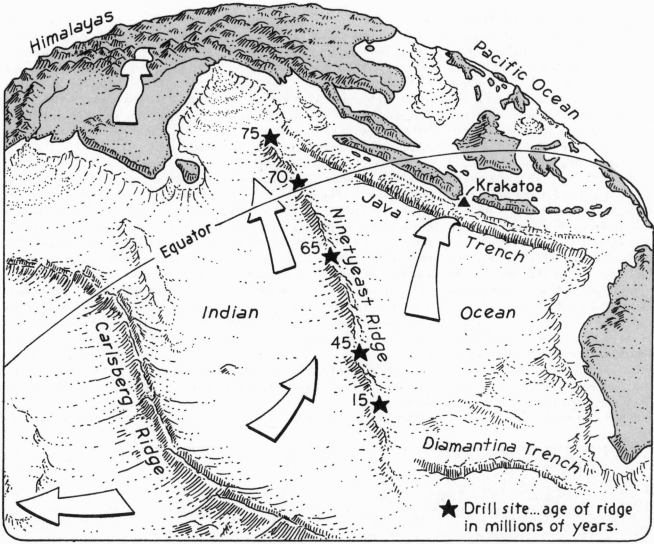

Himalayas

Pacific Ocean

Equator

Krakatoa

Java

Trench

Carlsberg Ridge

Ninetyeast Ridge

Indian

Ocean

75 ★
70 ★
65 ★
45 ★
15 ★

Diamantina Trench

★ Drill site...age of ridge
in millions of years.

cal samples (called "cores") of sediments from the ocean floor. Even though the crest of the ridge is deeply submerged along its entire length, the retrieved cores contained fossils of shallow-water organisms and low-grade coals and peats that must have been deposited at or near the surface. Each section of the ridge probably once stood as an island at the point where it was extruded. As this highstanding part of the mid-ocean ridge was carried away from the spreading center by plate motion, it gradually subsided to its present depth. The cores brought up from the five holes along the ridge confirmed this general hypothesis. The age of the sea floor deposits ranged from 75 million years at the

northern end of the ridge to 15 million years in the south.

The *Glomar Challenger* has made extraordinary contributions to our understanding of global geology. This remarkable vessel was designed specifically for scientific research. It could drill cores from the ocean floor in water depths as great as four miles. This task required the vessel to hold position within a few hundred feet directly above the bore hole for days at a time, although buffeted by wind and waves. This was accomplished by a dynamic computer-controlled positioning system linked to a sonar device on the ocean floor. Signals from the sonar device were received by hydrophones on the hull of the ship and

translated into pulses which controlled the ship's propellors and side thrusters. Only once, in the rough South Atlantic near the Falkland Islands, did the ship loose a "drill string" (the many lengths of steel pipe that drive the drill) because the positioning system failed to cope with wild weather. The ship's four-story laboratory was equipped for about twenty scientists and technicians. Cores could be analyzed on board or shipped to laboratories a-shore. The thousands of cores which were retrieved by this ship from the world's ocean floors constitute a rich library of the earth's past. The *Glomar Challenger* can fairly be said to have been the Palomar Telescope of geology.

49 Krakatoa

It is said that the sound of the concussion was heard in Madagascar, 3000 miles across the Indian Ocean. It was certainly heard as far away as Australia. It was the final convulsion of Krakatoa, a volcanic island in the Sundra Strait between Java and Sumatra. After three months of sporadic rumblings, the island came to full life violently on the afternoon of August 26, 1883, after a night of terrifying detonations and dusty darkness punctuated by flashes of brilliant light. On the morning of the 27th virtually the entire island disappeared with a mighty roar. Five cubic miles of material were blasted into the atmosphere from a magma chamber beneath the island. The explosion may have been partially powered by the expansion of steam that formed when sea water poured into the magma chamber through the ruptured walls of the volcano. Then the crust of the earth caved into the emptied chamber, creating a caldera, the kind of gaping water-filled basin we saw at Thera [36].

Unlike the Aegean island of Thera, Krakatoa was uninhabited. There were no towns or villages to be buried by ash or swallowed up by the sea. But we do have a vivid historical record of the effects of the massive tidal waves generated by the collapse of the volcano. Surprisingly, ships in the Sundra Strait safely rode out the long, rolling swells. The captain of the British ship *Charles Bal* recorded the scene in his log, and worried about the fate of the inhabitants of the coasts: "Such darkness and time of it few would conceive, and many, I dare say, would disbelieve. This ship, from truck to waterline, is as if cemented (by mud and ash); spars, sails, blocks and ropes in a terrible mess; but, thank God, nobody hurt or ship damaged. On the other hand, how fares it with Anjer, Merak, and the other little villages on the Java coast?" Where the tidal waves reached shallow water, and especially in the bays and inlets along the coasts of Sumatra and Java, they piled up into walls of water 100 feet high. The fragile villages of the coasts were no match for the mountains of water that broke over them. 36,000 people were drowned.

The catastrophe at Krakatoa was felt literally around the world. The shock wave of the explosion was recorded on pressure gauges worldwide. Sea waves registered on tidal gauges in the English Channel. But the most spectacular effects were a result of what the island's death did to the atmosphere. At least a cubic mile of fine volcanic dust was blasted into the stratosphere, twenty miles above the surface of the earth. These particles were so fine—a ten-thousandth of an inch or so in diameter—that they were able to stay aloft for years before settling to earth. The high velocity circulation of the upper

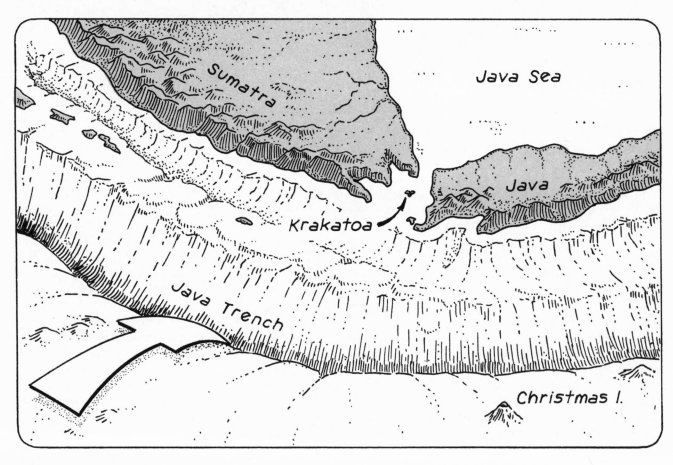

1882

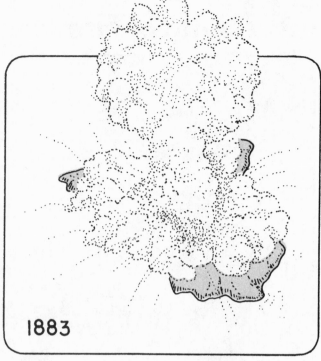

1883

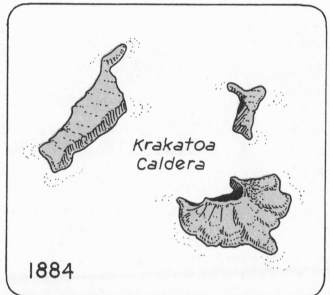

1884

Today

atmosphere quickly spread the particles into a broad equatorial belt that girded the planet. Within weeks this shroud of ash had drifted into mid-northern latitudes. Magnificent red sunsets became a nightly treat, and continued for months. It seems certain that this dust veil had a significant effect on climate. The amount of solar radiation reaching the lower atmosphere was decreased by as much as 20 percent during the first year following the eruption, and by 10 percent for several years thereafter. Average worldwide tempera-

tures were lower than normal.

The theory of plate tectonics provides a clear cause for the terrible eruption in the Sundra Strait. The island of Krakatoa is one of an arc of volcanic islands that stands just to the north of the Java Trench. The Java Trench is the result of a downward tug on the crust where the Indian Plate is being consumed by subduction into the earth's mantle. Energy generated along the descending plate heats the surrounding rock, which can rise toward the surface to form the sort of magma cham-

ber that emptied itself convulsively in 1883. It would seem that at Krakatoa the earth is seeking to fill the gap in the island wall—Sumatra and Java—which is breached by the Sundra Strait.

After the great eruption, Krakatoa remained at rest for forty-four years. Then in 1927 small submarine eruptions began in the old caldera. A small island was soon built up on the site of the crater. It is called Anak Krakatoa, or "Child of Krakatoa."

111

50 Ring of Fire

Mount Fuji...after Hiroshige, 1833

Where oceanic plates are consumed, volcanoes are born. Heat generated along diving plates stokes the fires of Mt. St. Helens, Vesuvius, Thera and Krakatoa. And nowhere are more volcanoes set alive by subducting plates than on the rim of the biggest plate of all. The floor of the Pacific Ocean consists almost entirely of one large lithospheric plate, the Pacific plate. This single piece of "eggshell" accounts for a sizable fraction of the earth's crust. Several smaller plates contribute to the floor of the ocean; the largest of these is the Nazca plate near South America. The Pacific Ocean is a shrinking ocean, pressed on all sides by advancing continents. The floor of the ocean is pushed back into the mantle at the trenches that rim its shores. And above the descending plates is a fiery necklace of volcanoes, the magnificent and deadly "Ring of Fire."

There are nearly 800 volcanoes known to have been active on earth during historic times. Five hundred of these are located on the perimeter of the Pacific. The most famous of the Pacific volcanoes is Fujiyama, or Mount Fuji, in Japan. Fuji rightly enjoys its fame. It is the tallest of that country's 200 active or inactive volcanoes. Over 12,000 feet high, it raises an almost perfectly symmetrical cone above the main Japanese island of Honshu, seventy miles west of Tokyo. Fuji is an imposing presence in the Japanese landscape, and in Japanese art and culture.

The union of delicate, symmetrical beauty with imposing strength and potential power has given Mount Fuji an important symbolic significance in the Japanese imagination. The mountain was long thought to be "the beginnings of heaven and earth, the pillar of the nation." Sacred Fuji is an object of veneration, and is climbed each year by many pilgrims from all parts of Japan. Shrines and temples dot its slopes.

Mount Fuji last erupted in the winter of 1707-8. The crowning crater, nearly a mile in diameter and 500 feet deep, is peaceful now, but the mountain remains a threat to the dense population that lives in its shadow. Fuji is an example of a strato-volcano, sometimes called a composite cone. It is the viscosity and gas content of the magma that rises in a volcano which primarily determines whether the eruption will be explosive or steady and deliberate. Basaltic lavas tend to flow freely and pour out of fissures in the crust to form low, broad "shield" volcanoes such as Mauna Loa in Hawaii, or continental sheets of lava such as we saw in the Pacific northwest, northern Ireland, and the Hudson River Palisades. On the other hand, if the molten material is thick and rises slowly, it will tend to harden in the throat of the volcano and the gases contained in the magma will be constrained from escaping. The trapped gas will build up a pressure that must ultimately be released by a violent detonation. The force of the detonation can blast

the hardened plug in the throat of the volcano into small fragments. In addition, molten rock blasted aloft will harden in the air. All of this particulate debris falls back to earth as cinder and ash, forming the tall symmetrical cones of volcanoes such as Fuji. The cinder layers are often reinforced by layers of lava from less explosive eruptions of the central cone or from cracks and fissures on the mountain's flank. It is the alternating layers of cinder, ash and lava that give a volcano such as Fuji its "composite" character.

Most of the famous volcanoes of

the earth's crust, such as Fuji and Vesuvius, are of the composite type and are usually associated with subducting plates. The symmetrical cones of composite volcanoes are everyone's idea of what a "volcano" should look like. The drawing at left, adapted from a painting of the nineteenth century artist Hiroshige, somewhat exaggerates the steepness, whiteness and symmetry of the actual Mount Fuji. It also conveys a peacefulness and purity of feeling that stands in stark contrast to the violence of that mountain's past—and possible future.

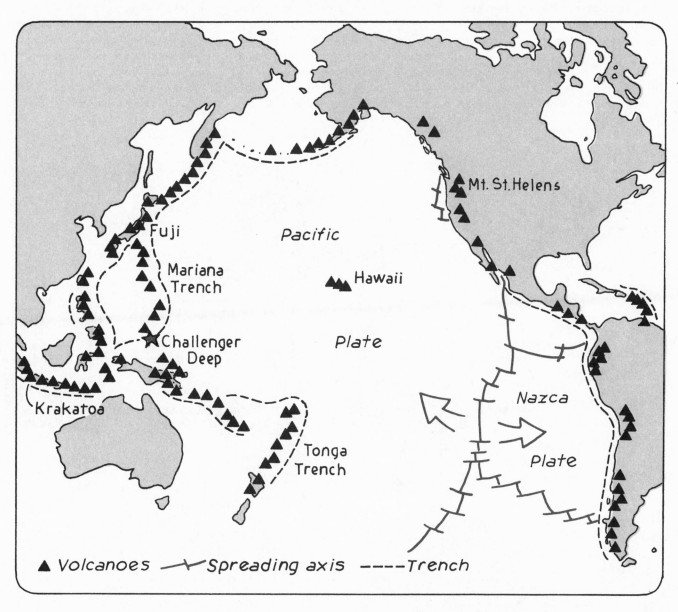

▲ Volcanoes ⌁ Spreading axis ----Trench

51 The Deepest Dive

If you held in your hands a carefully scaled relief model of the earth which was as large in diameter as this opened book is wide, then all of the surface irregularities, from the peak of the highest mountain to the floor of the deepest ocean trench, would occupy a layer not much thicker than this piece of paper. On a human scale, a bump on the earth's crust such as Mount Fuji or Mount Everest seems high indeed. But if you ran your finger lightly across the surface of our imaginary book-sized relief globe, you would scarcely feel any roughness. Shake the thin film of water from the globe's ocean basins, and still the exactly scaled model globe would be as spherical and smooth to the touch as a bowling ball.

It is the convergence of lithospheric plates that raises continental mountains and creates ocean trenches. New ranges of folded and faulted mountains, such as the Himalayas or the Alps, stand along boundaries where plates are pushing together. Old eroded mountain ranges, such as the Appalachians or the Urals, stand where continents have collided in the past. Caught in the squeeze of plates, continental rocks have nowhere to go but up. But where oceanic plates push against continents or against other oceanic plates, the ocean crust is forced back down into the mantle, generating trenches, earthquakes and volcanoes. Mountains are eroded even as they are pushed up. Trenches are filled with sediments even as they are pulled down. The height of continental mountains or the depth of oceanic trenches are a measure of the vigor of plate convergence.

The highest mountain on the earth's crust, Mount Everest in the Himalayas, was first climbed by Edmund Hillary and Tenzing Norgay in 1953. The deepest trough of the earth's crust, the Challenger Deep in the Pacific's Mariana Trench, was explored by Jacques Piccard and Don Walsh on January 23, 1960. The Challenger Deep lies about 200 miles south of Guam Island. It is part of the deep trench created by the downward thrust of the subducting Pacific Plate. The floor of the Challenger Deep lies seven miles below sea level, more than a mile deeper than Everest is high. No sunlight penetrates to this world of permanent inky darkness.

To make their plunge into this cold dark abyss, Piccard and Walsh used a remarkable vessel designed by Jacques Piccard's father, Auguste Piccard, the famed explorer of the deep sea and high stratosphere. That vessel, the *Trieste*, had to withstand

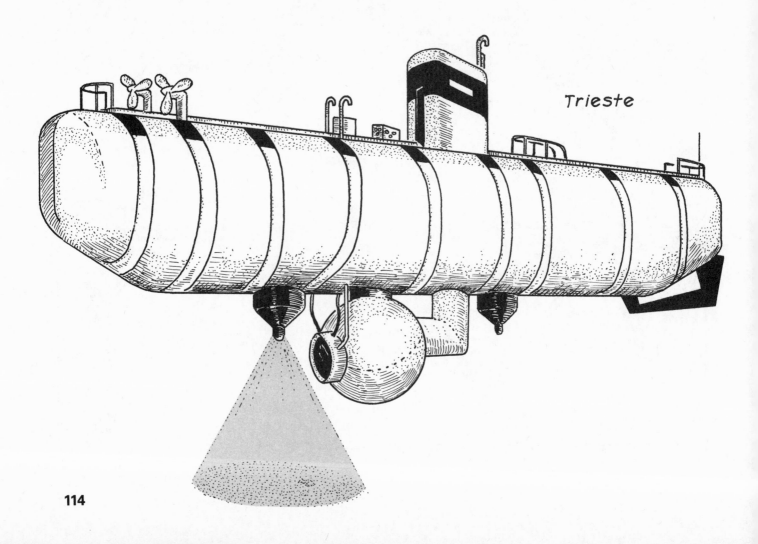

Trieste

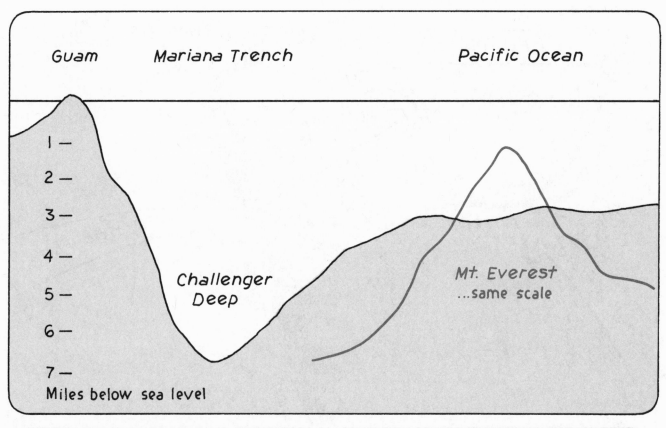

Guam   Mariana Trench   Pacific Ocean

1 —
2 —
3 —
4 —
5 —
6 —
7 —
Miles below sea level

Challenger Deep

Mt. Everest
...same scale

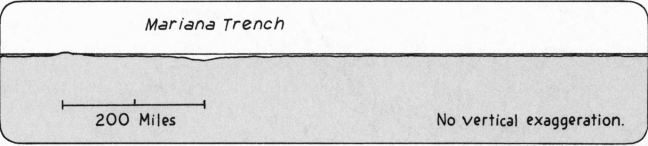

Mariana Trench

|← 200 Miles →|

No vertical exaggeration.

a pressure of eight tons per square inch and provide a life support system for two occupants suspended in watery space seven miles below the mother ship. What did Piccard and Walsh find on the floor of the trench? Certainly nothing resembling the rich variety of life forms sustained by the warm sunlit sea nearer to the surface, or like the strange communities of bottom life on the axis of the East Pacific Rise [1]. The floodlights of the *Trieste* disclosed a bottom of ivory-colored sediments. A shrimp. A fish, sole or flounder, somehow surviving the crushing weight of water. Slim pickings!

Beneath the thick sediments that partially fill the trench, the floor of the Pacific Ocean dives to destruction. This area of the Pacific has been visited by another vessel of discovery, the *Glomar Challenger* [48]. (The Challenger Deep and the *Glomar Challenger*, by the way, are both named for the famous British research vessel of the nineteenth century, *H.M.S. Challenger*.) Cores of sediments brought up by the *Glomar Challenger* from the sea floor just east of the Mariana Trench were of a greater age than cores from any other part of the Pacific. The portion of lithospheric plate that is currently being consumed at the Mariana Trench is 150 million years old. That slab of sediment-burdened crust was extruded in Jurassic times along a mid-ocean ridge thousands of miles to the east of the present trench. One hundred and fifty million years were required for the plate to move from the spreading axis in the east Pacific to the subduction trench near Guam. No piece of ocean floor anywhere on earth is more than a few hundred million years old. Unlike the very ancient continental rocks we visited in Canada and Greenland, the basaltic rocks of the ocean floors are continuously reprocessed through the mantle of the earth.

52 Down the Trench

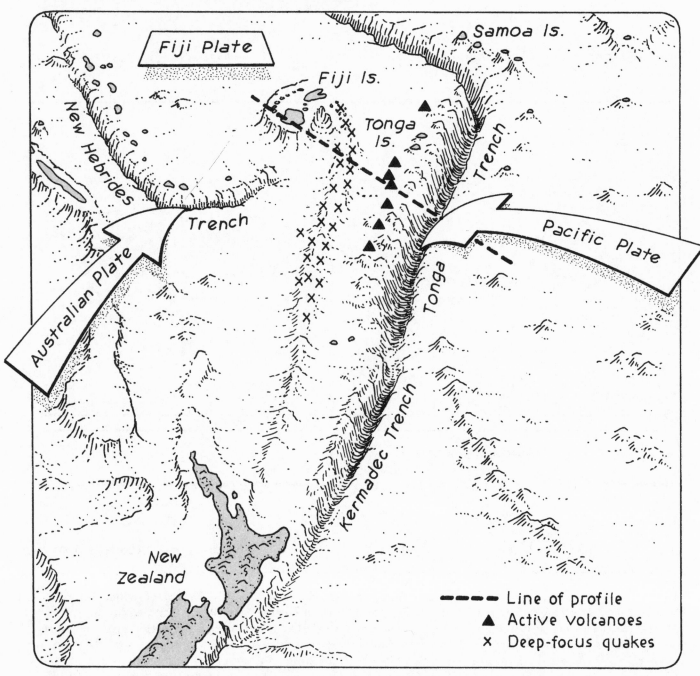

Fiji Plate

Samoa Is.

New Hebrides

Fiji Is.

Tonga Is.

Tonga Trench

Australian Plate

Trench

Pacific Plate

Tonga Trench

Kermadec Trench

New Zealand

- - - - Line of profile
▲ Active volcanoes
✕ Deep-focus quakes

What is the mysterious force that moves the earth's lithospheric plates? It is widely believed that convection currents in the asthenosphere exert a drag on the overlying plates which pulls them along. Or possibly the plates are pushed by rising magma at the mid-ocean ridges or at thermal plumes and "hot spots." Perhaps the plates are dragged away from the ridges by the weight of the cold subducting

plates sliding back into the mantle at convergent boundaries. Perhaps all of these forces are involved (see Introduction). The picture of the driving mechanism is not yet clear. But the creation, motion, and subduction of oceanic plates is a well-established fact. No part of the ocean floor anywhere on earth is older than 200 million years. Every 200 million years or so, ocean floors are created afresh at the mid-ocean ridges and

"gobbled up" at the oceanic trenches.

A visit to the Tonga Trench north of New Zealand offers an excellent opportunity to observe the destruction of lithospheric plate by subduction. Seismic activity in the area has been carefully surveyed, and the pattern of earthquake foci clearly reveals the disposition of the diving plate. Seismic activity near the Tonga Islands is confined to the re-

116

gion west of the trench. The depth of focus of the many earthquakes that jolt the region is directly proportional to the distance from the trench. The deepest earthquakes lie some 400 miles below the surface and at about the same distance west of the trench. A vertical reconstruction of the distribution of earthquake foci, as deduced from the seismic record, is shown on this page. They nicely outline the cold and rigid subducting plate that slides below the Tonga Islands at an angle of about 45 degrees.

All terrestrial earthquake activity occurs in the thin, rigid "eggshell" or lithosphere. Below the lithosphere, rock is too warm to break or shear with violence. Most earthquakes, such as the San Francisco, Yellowstone or New Madrid quakes [2,9,13], occur within a hundred miles of the surface. Wherever deeper earthquakes occur, they reveal the presence of rigid "eggshell" descending into the upper mantle. As lithospheric plates descend, thermal and mechanical stresses cause many small shearing shocks. The rapidity with which a plate descends determines the frequency and intensity of earthquake activity. Along the Tonga Trench the Pacific Plate slides below the Australian Plate at the rather "speedy" rate of three inches per year. At this rate the slowly warming plate remains sufficiently rigid to support earthquake activity even after it has pushed hundreds of miles into the earth. About 70 percent of the world's deep-focus earthquakes occur in the region of the Pacific between the Tonga and Fiji Islands. At last, of course, the plate is heated to the temperature of the surrounding mantle. With complete thermal assimilation, earthquake activity ceases. Where plates converge less rapidly, as where the Juan de Fuca Plate descends beneath North America [3], thermal assimilation is achieved at a shallower depth and no deep focus earthquakes are recorded.

Heat generated along diving plates is the source of volcanic activity on the inward side of trenches (Mt. St. Helens, Fuji, the island arcs of the Pacific and Caribbean). The Tonga Islands are a relatively new island arc, with a string of young volcanoes barely peaking above the surface of the ocean.

The Tonga Islanders, who live without skyscrapers, tunnels and bridges, are in little danger from the earthquake machine that grinds away beneath their feet. But further north in Japan, where Mount Fuji stands above a diving plate, a huge human population lives in constant peril. Since 1900 Japan has experienced more than twenty five earthquakes as strong as the quake that shook San Francisco in 1906. In the city of Tokyo, more humans live in daily peril from the quaking crust than anywhere else on earth. That city was last given a devastating jolt in 1923. In that year a combination of quake, fire and tidal wave killed over 150,000 persons and destroyed a large part of the city. Below the basements of Tokyo's skyscrapers, below the deep tubes of the subway lines, the floor of the Pacific Ocean burrows like a mole into the mantle. With each creak and groan of the descending slab the city shakes.

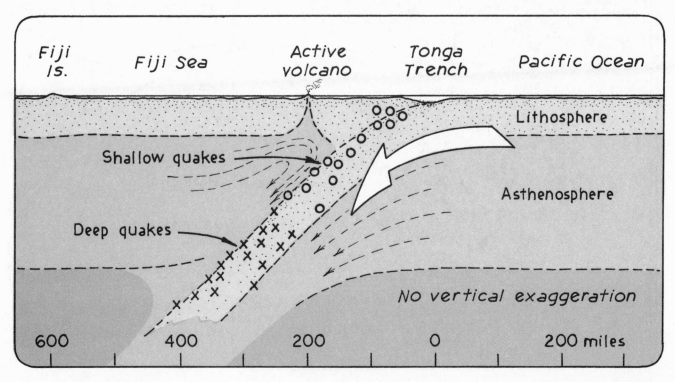

53 Great Barrier Reef

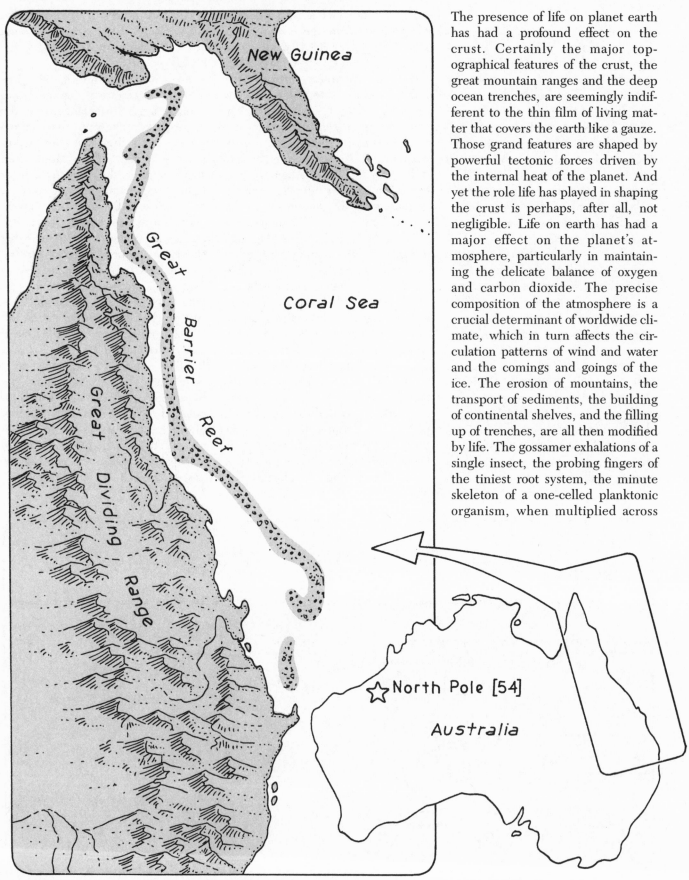

The presence of life on planet earth has had a profound effect on the crust. Certainly the major topographical features of the crust, the great mountain ranges and the deep ocean trenches, are seemingly indifferent to the thin film of living matter that covers the earth like a gauze. Those grand features are shaped by powerful tectonic forces driven by the internal heat of the planet. And yet the role life has played in shaping the crust is perhaps, after all, not negligible. Life on earth has had a major effect on the planet's atmosphere, particularly in maintaining the delicate balance of oxygen and carbon dioxide. The precise composition of the atmosphere is a crucial determinant of worldwide climate, which in turn affects the circulation patterns of wind and water and the comings and goings of the ice. The erosion of mountains, the transport of sediments, the building of continental shelves, and the filling up of trenches, are all then modified by life. The gossamer exhalations of a single insect, the probing fingers of the tiniest root system, the minute skeleton of a one-celled planktonic organism, when multiplied across

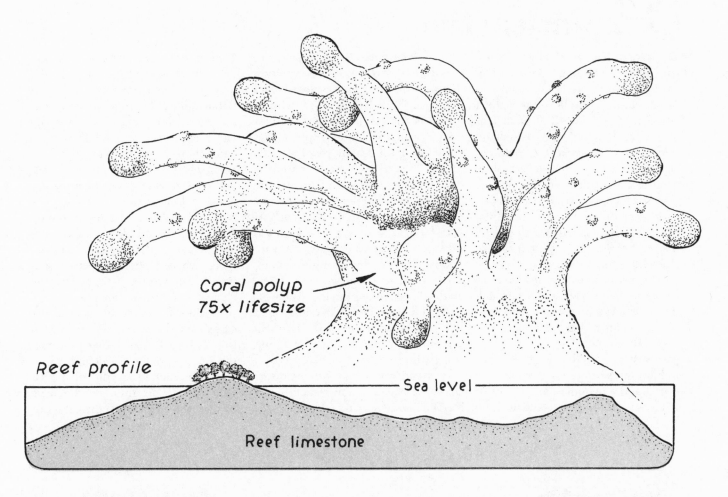

Coral polyp
75x lifesize

Reef profile

Sea level

Reef limestone

space and time by the myriad numbers of life on earth, can warm the planet, hurry mountains to the sea, or build up towering cliffs of chalk.

Perhaps the most convincing demonstration of the geological prowess of life is the Great Barrier Reef along the northeastern coast of Australia. The reef is a mighty limestone rampart, 1250 miles long, hundreds of feet thick, covering 80,000 square miles of the continental shelf. Against the Great Barrier Reef, the greatest constructive works of humans—a Great Wall of China or a Panama Canal—pale to insignificance. This colossal submarine "mountain range" has accumulated over the eons from the skeletons of animals, the gorgeous and varigated corals, cemented together by the limy secretions of algae. The reef faces the full power of the Pacific. Waves that have gathered strength across thousands of miles of open water hurl themselves against this

shore. If the reef were made of granite it would have long since crumbled and been ground to sand. But life is resilient and persistent. The reef is alive. What the waves destroy, life patiently rebuilds.

The forms and colors of the living corals are astonishing: fans, ferns, brains, trees, flowers, and antlers of a paintbox brilliance. A typical coral polyp (*pocillopora damicornus*) is shown above greatly magnified, looming above a cross-sectional drawing of a coral reef. The creature is tiny, only a fraction of an inch in diameter. Its twelve ghostly tentacles, lined with golden algae, wave about in search of food. Spring-loaded stinging mechanisms along each arm wait to disable a passing victim. Mucus secreted by the tentacles helps snare the victim and pass it along to the central mouth. At its base the creature builds a tiny, anchoring castle of limestone, delicately fluted, as frail as eggshell. Of such stuff has the massive reef been made.

On the seaward wall of the Great Barrier Reef there are ribbon reefs, several miles long and seldom more than a quarter mile wide. Behind this protecting rampart lie irregular patch reefs, ring-shaped atolls and islands of wave-crushed coral sand and boulders. Within the confines of this fairyland environment, life labors to build earth's crust. Reef systems of past eons have contributed massive geological formations. Ancient fossil reefs can be found in such unlikely places as Tennessee, the Canadian Rockies, and the Arctic!

Human civilization, with its attendant pollution, encroaches on the coral domain in the guise of boaters, divers, anglers, photographers and collectors. In spite of its ability to survive pounding ocean breakers, the reef system depends for its resilience on a delicate balance of environmental factors. It will be an easy victim of careless human intervention.

54 Ancient Life

The Great Barrier Reef of eastern Australia is a teeming habitat for life. An astonishing variety of creatures—plants and animals—make their home in those beautiful coral gardens. Two thousand miles west of the reef, all the way across the continent of Australia at a hot desert place paradoxically called North Pole (see map page 120), Australian geologists have reconstructed an earlier habitat of life, and have given us a glimpse of the crust of the earth as it might have looked 3.5 billion years ago.

The fossil record of macroscopic life forms goes back about 650 million years. The first multicelled creatures, the invertebrates, appeared at about that time. The fish and land plants were on the scene by 400 million years before the present. The reptiles, the mammals and the birds followed within the next few hun-

dred million years. And a few more hundred million years of evolution led to *homo sapiens*. Geologists and biologists have long wanted to know when in the earth's history the first living forms of matter appeared on the planet. In recent years increasing evidence has accumulated to suggest that single-celled organisms, bacteria and blue-green algae, had evolved in the earth's seas as early as 3.5 billion years ago. Even these simple creatures, almost miraculously preserved as microfossils, show wonderful variations of form, with shapes like threads, spheres, stars and parachutes. At a time when the land areas of the earth were still completely barren, indeed at a time when the continental crust was still young and warm, marine organisms already thrived in abundance.

The evidence for life at North

Pole in western Australia, although not entirely unambiguous, is in the form of fossil structures called stromatolites. These moundlike constructions were very common in the late Precambrian era, a billion years ago, when stromatolite communities colonized widespread areas of shallow seas. At that time, the algae did not yet have multicelled competitors. Stromatolites are less common today, but are still deposited in shallow water and intertidal environments where exposed surfaces of sediment are covered with a layer of algae. The algae form a mat of threadlike filaments lying near the surface of the water. When a high tide or rough water sweeps across the mat, particles of calcium carbonate are trapped by the filaments. Quickly, seeking sunlight, the algae extend thin filaments through this

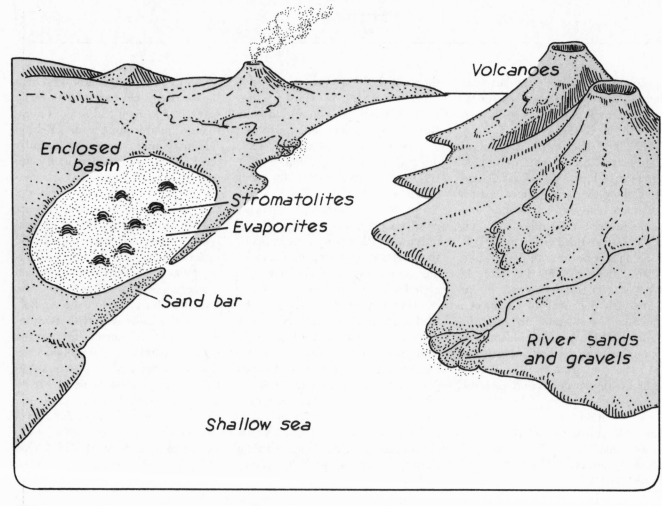

Volcanoes

Enclosed basin

Stromatolites

Evaporites

Sand bar

River sands and gravels

Shallow sea

fresh layer of particles, binding it into the growing structure and re-populating the top surface. The next high water brings a new layer of particles. In this way, pillars or pillows of finely layered carbonates are built up. Structures closely resembling present day stromatolites have been found worldwide in sedimentary rocks (often highly metamorphosed) billions of years old.

Few sedimentary rocks more than a billion years old have escaped drastic modification by the powerful tectonic forces which are constantly at work reshaping the crust of the earth. The 3.5 billion year old sedimentary sequence at North Pole is untypical in that it shows little evidence of metamorphism. The rocks at North Pole have evidently avoided the conditions of heat, pressure and extensive folding that has modified most other sediments of such great antiquity.

Careful geological detective work has revealed the nature of the lavas, sands, silts and evaporites which were the parent components of the North Pole rocks. From the sequence of basalts and sediments the early habitat for life has been reconstructed. The interpreted sequence of sediments is illustrated schematically at right. An imaginative reconstruction of the kind of environment in which these sediments were deposited is shown at left. In this ancient habitat blue-green algae apparently flourished and built domelike structures from particles snared from the sea.

The North Pole stromatolites, if that is what they are, are much simpler than the elaborate columnar and branching forms found in some younger rocks, or those built up today by contemporary algae architects. The layered structures at North Pole may yet turn out to have been caused by a non-living agency. More likely, they give a hazy picture of the very origins of life on earth.

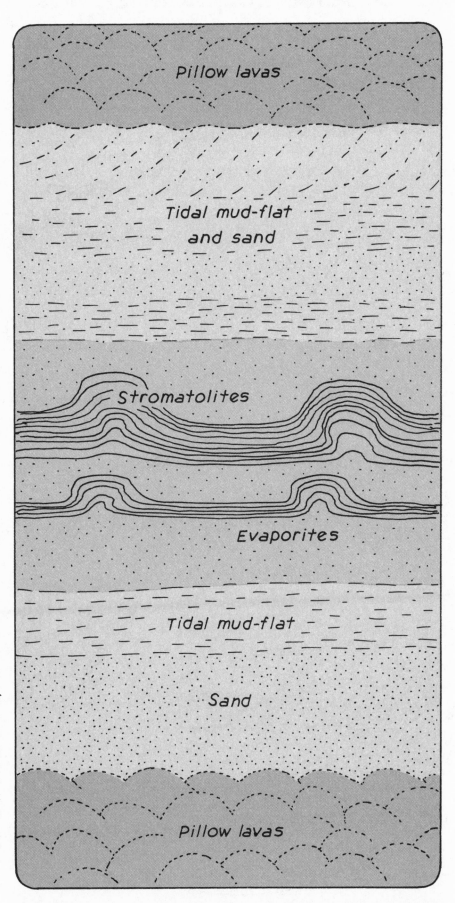

55 Assembly Line Islands

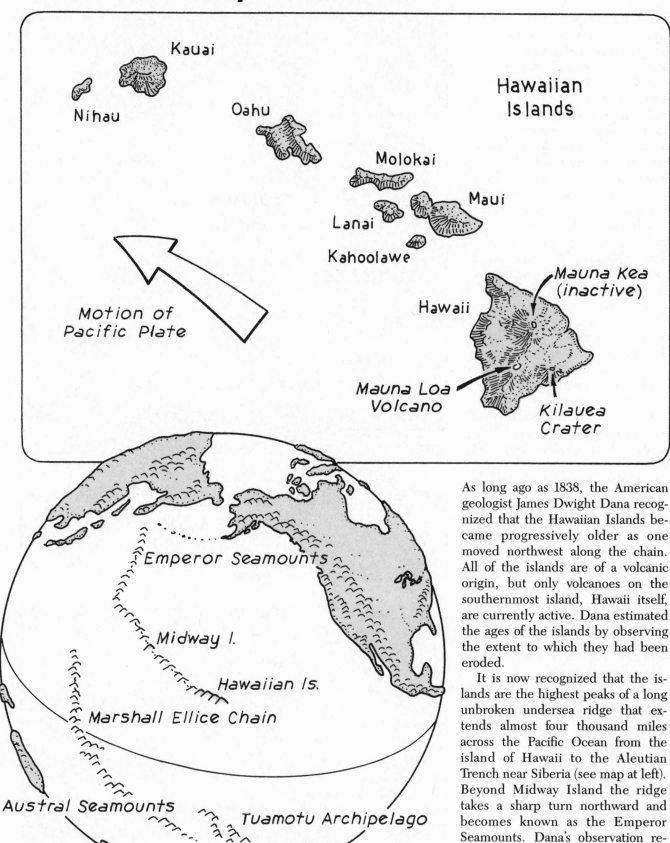

Kauai

Nihau

Oahu

Molokai

Maui

Lanai

Kahoolawe

Hawaiian Islands

Mauna Kea (inactive)

Hawaii

Mauna Loa Volcano

Kilauea Crater

Motion of Pacific Plate

Emperor Seamounts

Midway I.

Hawaiian Is.

Marshall Ellice Chain

Austral Seamounts

Tuamotu Archipelago

As long ago as 1838, the American geologist James Dwight Dana recognized that the Hawaiian Islands became progressively older as one moved northwest along the chain. All of the islands are of a volcanic origin, but only volcanoes on the southernmost island, Hawaii itself, are currently active. Dana estimated the ages of the islands by observing the extent to which they had been eroded.

It is now recognized that the islands are the highest peaks of a long unbroken undersea ridge that extends almost four thousand miles across the Pacific Ocean from the island of Hawaii to the Aleutian Trench near Siberia (see map at left). Beyond Midway Island the ridge takes a sharp turn northward and becomes known as the Emperor Seamounts. Dana's observation regarding the ages of the islands seems to apply to the entire angled chain of volcanic seamounts. The northernmost seamounts in the Emperor chain appear to be about 70 million years old. Koko Seamount, just north of the bend, has been reliably dated at 46 million years. Midway Island is about 18 million years old. And, of course, the island of Hawaii is still under construction by the fiery instruments of volcanism.

In 1963 the geologist J. Tuzo Wilson suggested an explanation for this curious progression of ages that related the origin of the islands to the newly emerging theory of plate tectonics. According to Wilson, the islands were produced successively, assembly line fashion, as the rigid Pacific plate moved northwest over a source of magma anchored in the asthenosphere. Wilson called that source of volcanic material a "hot spot." The hot spot is located beneath Kilauea Crater on the island of Hawaii and supplies the lava that builds that peak.

The slow drift of the Pacific plate will carry Hawaii, like its older siblings, away from the hot spot. The magma will find a new vent to the surface and begin building a new island. Meanwhile, erosion will cut down the high cones of Mauna Kea and Mauna Loa.

Many lines of reasoning confirm Wilson's idea. Other chains of Pacific seamounts lie parallel to the present motion of the Pacific plate and invite a similar explanation (the motion of the plate seems to have changed direction about 40 million years ago). The ages of the Hawaiian Islands and their distances from the active volcanoes of Hawaii are roughly consistent with the rate of motion of the Pacific plate deduced from magnetic anomalies [25]. Furthermore, the inclination to the horizontal [47] of fossil magnetism in the volcanic basalts of Midway Island suggests that when the island formed it was closer to the equator than presently, at a latitude not significantly different from that of Kilauea Crater.

The Hawaiian hot spot is only one of many such thermal features which geologists have mapped on our dynamic planet. Over 100 localized regions of intense volcanic activity have been designated "hot spots." We have seen others at Yellowstone [9] and Iceland [23]. Unlike most of the earth's active volcanoes, these hot spots are not always located on plate boundaries. The Hawaiian hot spot is thousands of miles from the nearest plate boundary. The source of heat which sustains these mid-plate hot spots and the mechanism which anchors them in the upper mantle are still matters of speculation.

The Hawaiian Islands continue to ride the plate westward the way a surfer rides the waves. Tens of millions of years from now a longer journey will be required for tourists to reach the island of Hawaii from the continental United States. But by that time the island will have eroded beneath the waves, and, if Wilson is right, new islands will have emerged above the Hawaiian hot spot to attract the tourists.

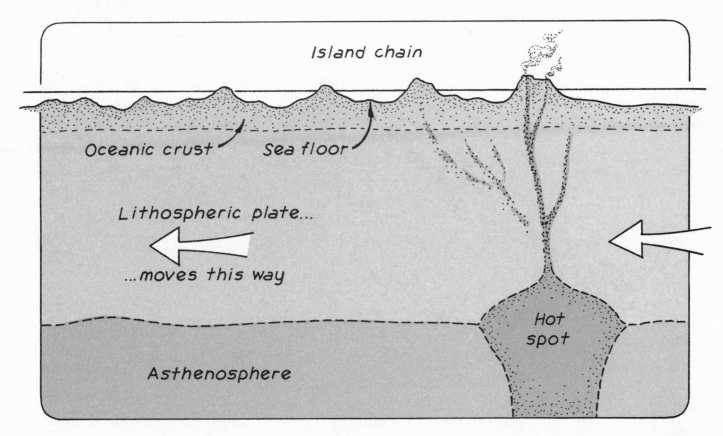

Island chain

Oceanic crust — Sea floor

Lithospheric plate...

...moves this way

Hot spot

Asthenosphere

56 The Panama Connection

The Isthmus of Panama is an unlikely link between two great continents. It is a connecting thread so slender that the wonder is that it is there at all. For thousands of years the link was a path for human migration between north and south. Later generations of sea captains considered the isthmus an annoying wall between two oceans, a barrier that often necessitated a detour of thousands of miles around Cape Horn. Breaching the barrier with a canal was one of the great engineering achievements of our century. Today, this thin wrinkle of the earth's crust, this curious quirk of geography, remains both a bridge and an obstacle, the strategic nexus of the Americas.

Recent evidence from the bottom of the sea suggests that the thin string of land connecting North and South America is, geologically speaking, a rather recent development. Only 4 million years ago the two great continents of the western hemisphere were separated by a seaway, and a ship could have sailed unimpeded from Atlantic to Pacific.

Paleobiologists who studied land fossils have long recognized this fact. The fossil record on the two continents shows a divergence of flora and fauna during the many tens of millions of years that followed the breakup of Pangaea. Then, beginning several million years ago, a convergence of fossils indicates the establishment of the Panama bridge. The South American armadillo, for example, made its way north and established itself in Central America and the southern United States. A greater traffic went the other way. The mastodon, the tapir, and several kinds of carnivores moved south. Southern species fared worst in the exchange that took place across the new bridge. Many South American species that had flourished in splendid isolation were unable to compete with the northern invaders and became extinct.

Sediment cores [48] brought up from the ocean floor on opposite sides of the isthmus tell the same story from a different perspective. Fossil skeletons of one-celled marine organisms called foraminifers, buried in sea floor sediments, began to exhibit diverging forms at about the same time land fossils began the trend toward convergence. This presumably resulted from the establishment of two separate faunal provinces, which then evolved in isolation along different paths.

The Isthmus of Panama seems to have begun its rise from the sea about 20 million years ago as spreading from Pacific rift zones pushed sea floor against Central America. Ocean crust was consumed by subduction along the Middle America Trench, parallel to the present western coasts of Costa Rica and Panama (see map page 000). The subducting plate pushed up the overlying crust

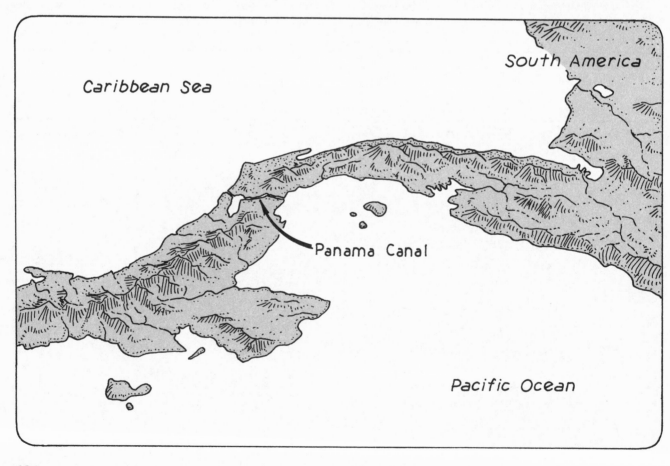

Caribbean Sea

South America

Panama Canal

Pacific Ocean

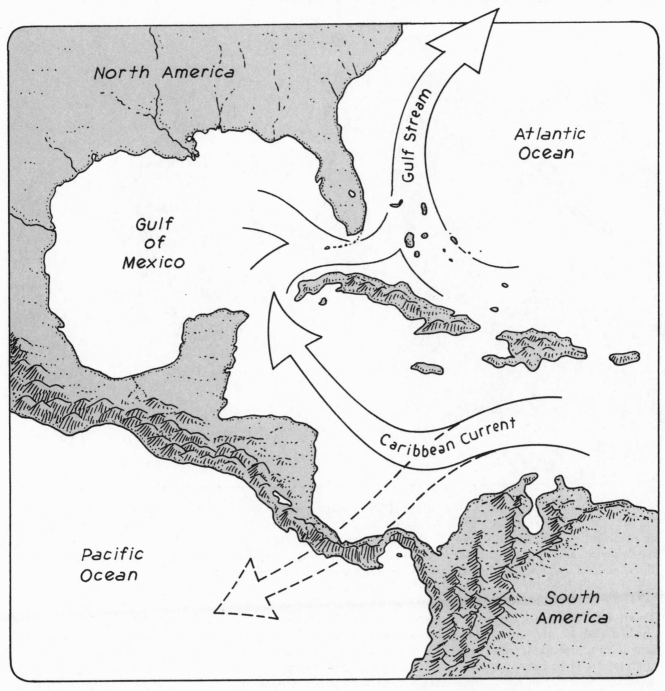

and injected into it volcanic lavas. By about four million years ago a continuous barrier had been raised between the two oceans and the continents were linked.

The closing of the Panamanian seaway may have had even more dramatic effects than have yet been described. The last great series of ice ages in the northern hemisphere began at about the same time the connecting ridge was pushed up. Before the raising of the isthmus, the waters of the two oceans intermingled. In particular, a warm westward-flowing Atlantic current flowed into the Pacific (dashed arrow on map above). When the Panama dike was established this current was deflected northward to strengthen the Gulf Stream. A more powerful Gulf Stream brought more warm moist air into the northern latitudes and increased the level of precipitation.

One might think such a development might bring about the end of a regime of ice, not initiate one. But it is moisture that is needed to grow ice. When more winter snows fell in central Canada and northern Europe than melted in summer, a new ice age began. In this way a squeeze of crustal plates in tropic latitudes may have helped initiate the growth of mighty continental ice sheets on North America and Europe!

57 Fossils in High Places

High Andean peaks and Darwin's fossils

In the southern autumn of 1835, Charles Darwin, twenty-five-year-old naturalist on His Majesty's Ship *Beagle*, went ashore at Valparaiso, Chile, and set out on a journey of several weeks by mule across the backbone of the Andes Mountains. Darwin's Chilean hosts were astonished that a young man should come halfway around the world from England to walk around in the mountains, hammer in hand, dropping chips of rock into collecting bags. Darwin replied by asking if they were not curious to know how and why earthquakes and volcanoes occurred, and why some springs were hot and others cold. Along the coast of Chile Darwin himself experienced a dramatic earthquake and an exhibition of volcanic pyrotechnics. And while travelling in the high Andes he began to piece together a tentative picture of the long violent history of the western coast of South America.

At an elevation of 12,000 feet Darwin found fossil seashells, embedded in sedimentary rock formations that could only have been laid down on the floor of an ancient sea. That was puzzle enough, but more was to come. High on the Argentine side of the range he came upon a forest of petrified trees, embedded in a thick matrix of sandstone. Clearly (to the curious young naturalist) the land on which these trees grew must have subsequently been depressed so as to have been covered by the sea. Only in that way could the trunks of the trees have been encased in a matrix of sand. Then the sand-entombed trees were pushed vertically upward 7000 feet above sea level, to the place where he found them. Such a sequence of events must have seemed improbable, even impossible, at a time when most people

believed the earth was only a few thousand years old. And yet, how else could Darwin account for what his geological hammer revealed? The entire range of the Andes, Darwin realized, had been pushed up out of the sea by some tremendous force. "Nothing, not even the wind that blows," he wrote, "is so unstable as the level of the crust of the earth."

Darwin knew that the raising of the mountains was not a single catastrophic event, as Captian Fitzroy of the *Beagle* would have argued, an event associated with the biblical Creation or the flood of Noah. Rather, the Andes were the result of a continuous pressure acting over eons of geologic time. Early in 1835 Darwin visited the Chilean town of Concepción shortly after it had been demolished by a terrible earthquake. The land around the town's harbor had been raised several feet with respect to

the level of the sea. This was the
agency of elevation which took root in
Darwin's imagination. Foot by foot,
inch by inch, given enough time, so
had the towering Andes been lifted
from the sea.

But what was the source of the
force that shook the earth and raised
the mountains? And why was that
force so active along the western coast
of South America and not, for example, equally efficacious in the quiet
green midlands of faraway England?
Darwin did not know. Only with the
advent of the theory of plate tectonics
has the full story of Andean mountain-
building been made clear. Along the
western flank of South America, the
Nazca plate, part of the floor of the
Pacific Ocean, is being thrust beneath
the continental crust (see map page
000). It was the pressure of that subducting plate that crumpled up the
Andes Mountains, lifted fossils from
the sea, wrecked the town of Concepción, and stoked the fires of the
volcano Osorno which Darwin
watched in fiery eruption in January of
1835.

The Peru-Chile Trench marks one
of the longest subduction zones on
earth. It is also one of the most active
arenas of plate consumption. The floor
of the Pacific dives beneath South
America at a rate of almost four inches
per year, with terrible effect on the
land above. A vivid example of the
work of the subducting plate was the
Peruvian earthquake of May 31, 1970,
the deadliest in the recorded history of
the South American coast. Fifty-
thousand people died, and nearly a
tenth of the population of that country
were left homeless. Darwin knew
nothing of subducting plates, but as he
crossed the Andes he knew the earth
was at work beneath his feet, and that
it had been at work for many more
than the 6000 years of earth history
accepted by Captain Fitzroy. Not even
the wind moves, Darwin knew, with
so steady and unrelenting a force.

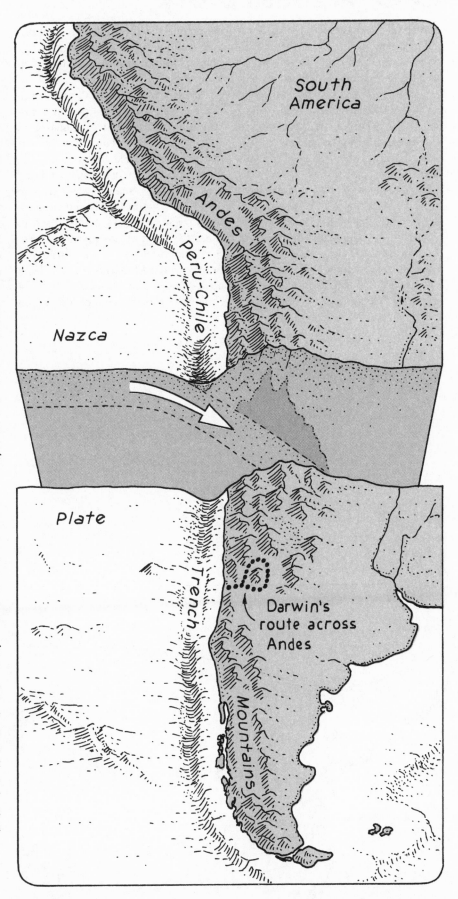

58 A Fossil Story

The small lizard-shaped reptile illustrated below has its own story to tell about continental drift. The resourceful Mesosaurus was less than two feet long, and its long jaws and needlelike teeth seem to have been ideally suited for sieving tiny crustaceans from brackish coastal waters and river estuaries. It was probably one of the earliest of the land-living reptiles to readapt to life in the water. It may have been an ancestor of the completely aquatic dinosaurs, the ichthyosaurs, which prowled the seas at the climax of the Age of Dinosaurs. The fossils of Mesosaurus are found in mudstones which were deposited in the Permian era, some 250 million years ago. They

are found in eastern South America and southwestern Africa, but in no other place on earth (see bottom map at right).

Alfred Wegener, the father of continental drift, used the puzzling distribution of the fossil Mesosaurus to support his theory of an ancient supercontinent called Pangaea ("all-earth"). No one in Wegener's time supposed that this little shallow-water reptile could have swum 2000 miles across a deep ocean basin to establish itself in separate provinces on opposite sides of the Atlantic. Some paleontologists proposed that Mesosaurus crossed the ocean on a land bridge that once connected the two continents, a land

bridge that subsequently must have sunk into the ocean floor. Wegener dismissed the idea of a land bridge as impossible. The continents stand higher than the sea floors, he said, because they are composed of light granitic rocks which float on the denser basalts of the upper mantle like blocks of ice in water. It would be inconceivable, he maintained, for lighter rocks to sink into heavier ones. Wegener suggested that it was horizontal movements of the earth's crust, not vertical displacements, that accounted for the puzzling distribution of the little swimming reptile.

At the time Mesosaurus thrived, said Wegener, South America and

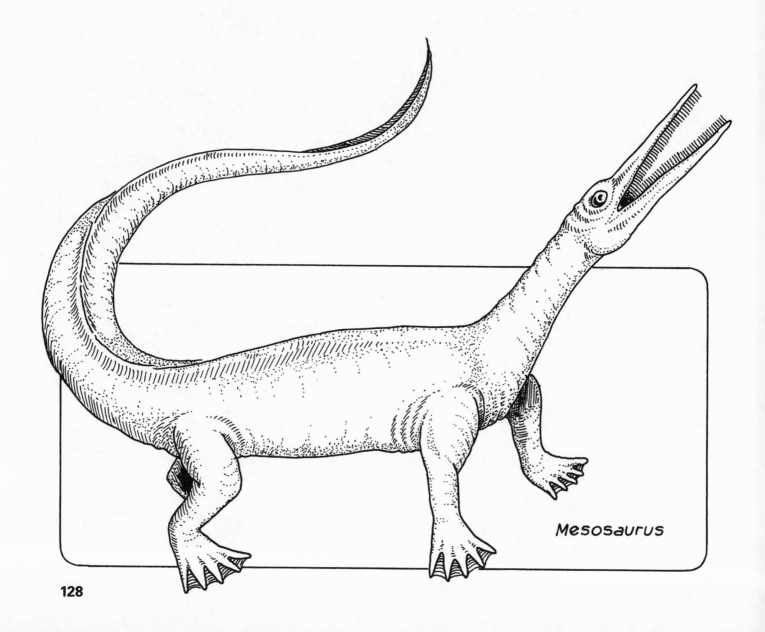

Mesosaurus

Africa were one continent, covered in places with shallow inland seas (see top map). Mesosaurus did not swim the deep ocean; rather, the breaking up of Pangaea some 140 million years ago and the subsequent drift of continents carried the fossil remains to their present locations.

Mesosaurus is only one of many fossil plants and animals that point to the existence of an ancient supercontinent. Another distinctive collection of fossil plants common to 200 million year old rocks of the southern continents is known as the Glossopteris flora, after a prominent species in the assemblage. *Glossopteris* was a large seed fern which flourished over wide low swampy areas near the margins of glaciers. The fossil ferns are common to southern South America, southern Africa, India, Australia and Antarctica. Proponents of shifting continents have argued that it would be unlikely for this diverse assemblage of complex plants to have developed in identical ways in such diverse environments as India and Antarctica.

Wegener used these ancient fossil distributions, along with present-day distributions of flora and fauna and corresponding geological features on opposite sides of the Atlantic, to support his theory of an ancient supercontinent. Summarizing his arguments, he wrote: "It is just as if we were to refit the torn pieces of a newspaper by matching their edges and then check whether the lines of print run smoothly across. If they do there is nothing left but to conclude that the pieces were in fact joined in this way." The argument sounds persuasive today, but as we have seen elsewhere, Wegener's fellow scientists remained unconvinced that the massive continents could barge around on a supposedly rigid earth. Fifty years would pass before Wegener's ideas would find wide acceptance.

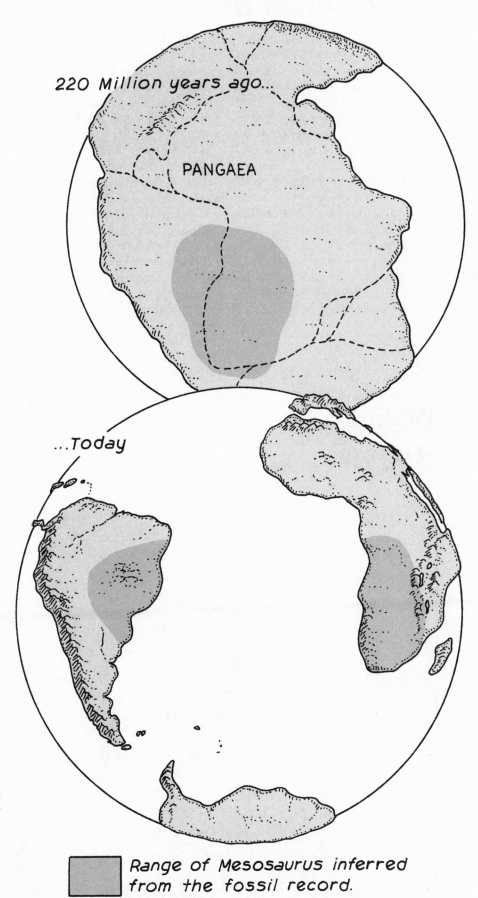

Range of Mesosaurus inferred from the fossil record.

59 Antarctica Under the Ice

The heroic days of Antarctic exploration are past. Nowadays scientists reside year-round on that icebound continent, snug in prefabricated villages buried beneath the snow. Travel is by helicopter, Snow-cat or Caterpillar tractor. Nuclear power keeps the lights burning. But earlier in this century, when Shackleton, Admunsen and Scott blazed the way to the pole, every step of the journey was a test of human ability to endure the worst this planet had to offer. But even as they fought for survival, those heroic explorers of the southernmost continent found time for geology. Scott and his companions managed to reach the southern pole (barely beaten to that

elusive prize by Amundsen), but perished on the return journey to the coast. When, the following summer, searchers found the frozen bodies of Scott's party in their tents on the ice, a nearby sledge was still loaded with thirty-five pounds of fossil-bearing rocks. The rocks, including coal, contained imprints of the same warm-weather forms of vegetation that had also thrived in India, Africa and South America 200 to 300 million years ago.

And so was posed the riddle of Antarctica. How had steamy coal forests flourished where today not a single blade of grass will grow? Why should the fossil flora of the white continent at the bottom of the world have so much in common with the fossils of the tropical Gondwana province of distant India? The riddle deepened with the Antarctic explorations that began during the International Geophysical Year of 1957-58. Since that time geologists from several nations have collaborated on a detailed

reconnaisance of the frozen continent. In 1969 a rich lode of fossils was discovered at a place called Coalsack Bluff. Thousands of feet above sea level, scientists worked in sub-zero temperatures to extract from the crumpled strata evidence of life on earth during the Age of Reptiles. This remarkable natural museum of ancient life at once deepened the riddle and supplied the answer. The answer was a dramatic confirmation of the theory of continental drift. The Coalsack Bluff fossils were one of the keys that enabled geologists to confidently reassemble the continental jigsaw puzzle. Antarctica, it was decided, had only recently (geologically speaking) made its way to the earth's cold underside, drifting away from the other continents to which it was once joined in a more temperate climate.

Geologists are beginning to appreciate how the presence of a huge landmass near the southern pole might affect worldwide climate. Antarctica and Australia parted company about 50 million years ago as the southern continent began its slide toward the pole. Thirty million years ago, as the climate cooled, the mountains of Antarctica spawned widespread glaciers. Something resembling the present continent-spanning ice cap was in place 10 or 20 million years ago. Ice reflects back into space three or four times more of the sun's radiation than does vegetation. There seems little doubt that these past tens of millions of years have been significantly cooler than other eras of the earth's history, and the Antarctic ice cap may be part of the cause. Continental drift may be one of the most important regulators of the long term climate of the planet.

Antarctica presently carries a burden of over 7 million cubic miles of ice, 90 percent of the ice on the crust of the earth. In some places the Antarctic ice sheet is two-and-a-half miles thick. As in Greenland, the weight of this frozen blanket has depressed the

Coalsack Bluff ▲

● South Pole

Antarctica

Pacific Basin

South America

Atlantic Basin

Indian Basin

Africa

———— Limit of
ice sheet

continent. In places the surface of the mostly hidden continent has been pushed below the level of the sea. Strip away the ice from Antarctica and it would look something like the ice-free, broadly-flooded continent shown on the map above. Given enough time, however, the depressed crust would rebound, tipping the waters of the shallow inland seas back into the ocean basins.

But it would not be possible to strip the ice from Antarctica without putting the water someplace else. If a warming climate caused the southern ice cap to melt, the level of the world's oceans would rise as much as 250 feet, or about to the level of Lady Liberty's nose. The great coastal cities of the earth would be completely submerged!

131

60 Of Clams and Penguins (A Postscript)

We began this geological journey on the crest of the East Pacific Rise, miles beneath the sea, where tube worms and giant clams gleaned a sunless living from the earth's internal heat. We end on the Antarctic ice cap, in an environment almost equally hostile to life. Only a few flowering plants survive on the frozen continent, and those are on the Palmer Peninsula which stretches a lean cold arm toward the warmer climes of South America. Algae, lichens and mosses do better, but only barely. The lichens survive by sheer tenacity, the mosses by latching into whatever thin soil the lichens create.

Given the impoverished Antarctic flora, it is not surprizing that fewer than a hundred species of animals live south of the Antarctic Circle. Half of these are insects. There are no land mammals native to Antarctica. A few species of seals and whales inhabit the coastal waters. Birds are the most successful of the polar inhabitants, and the most imposing of these are the penguins.

The penguin *is* Antarctica, or was before the coming of humans. The sleek, flightless bird is wonderfully adapted to the harsh Antarctic environment. Not the least of its successful adaptations is a loyalty to its

young that is little short of astonishing. Penguins stay put on their nests through the long period of incubation, cold-shouldering blizzards and raging gales, warming the delicate eggs with the heat of their bodies. Refusing to abandon the incubating chicks even to seek food, the parent lives for weeks off body fat.

To this sparse yet beautiful land humans have come at last, recently in substantial numbers. It is fair to say that the majority of visitors to Antarctica have been persons—explorers and scientists—with a kindly regard for the environment. The continent may not be so lucky in the future. Humans have become a force for change on the crust of the earth as powerful in its effect as the deep mantle currents that drive the plates. But the time scale of the change is different. A delicate and beautiful environment that took moving plates 100 million years to fashion, can be tarnished in months by human carelessness. The potential for tragedy is great indeed. The unrestrained burning of fossil fuels could have a catastrophic effect on the atmosphere. It is not inconceivable that destruction of the ozone layer could expose life on the surface to dangerously increased ultraviolet radiation from the sun. Or that modification of climate could melt the polar ice caps and cause the seas to overflood the coastal lowlands. But these frightening long-term scenarios pale before the possibilities for destruction inherent in current stockpiles of nuclear weapons. The explosions of Mt. St. Helens and Krakatoa, the shaking earth at San Francisco and New Madrid, the greatest potential, in other words, the moving plates have for destruction, would seem as nothing compared to the firestorm of devastation unleashed by a nuclear war. We must foster the same solicitous care for the crust of the earth and for life on it as the penguin has for its chicks.

Emperor penguin and chick

Glossary

ASTHENOSPHERE. A layer of the earth's upper mantle, roughly between 50 and 200 miles below the surface, which is close to its melting point. The layer is less rigid than the rock above and below, and may be in convective motion.

AUSTRALOPITHECUS. Primates of the period 6 to 2 million years before the present with skeletal characteristics intermediate between apes and true humans.

BASALT. A finely grained, dark, dense volcanic rock. The primary constituent of the ocean floors.

BRACHIOPOD. Double-shelled marine invertebrates. More common in the Paleozoic than present.

CALDERA. A large basin-shaped depression or crater, caused by the inward collapse of a volcano after eruption.

CALEDONIAN. A mountain-building episode in northern Europe about 450 million years ago.

CATASTROPHISM. The belief that geologic history occurs as a sequence of sudden violent events, rather than slow, imperceptible increments of change. (See *uniformitarianism*.)

CENOZOIC. Literally, "recent-life." Refers to the period between 65 million years ago and the present.

COCCOLITH. A microscopic calcium carbonate disk-shaped covering secreted by a marine planktonic organism.

COMPOSITE (VOLCANO). Volcano formed by alternate eruptions of ash or cinders and liquid lava. Most have a symmetrical cone-shaped appearance. Also called a strato-volcano

CONTINENT. A slab of rock of granitelike composition which floats on the denser rocks of the earth's upper layers.

CONVECTION. A circular vertical movement of a fluid medium due to heating from below. Hotter, therefore less dense, material rises; cooler, heavier material sinks. Effective as an agent of heat transfer

CONVERGENT PLATE BOUNDARY. A boundary between lithospheric plates where the plates move together. In some cases an ocean plate might be consumed by subduction into the mantle, or continental masses may converge and collide

CORAL. Any of a large group of shallow-water, bottom-dwelling marine invertebrates which secrete calcium carbonate skeletons. Reef-building colonies common in warm waters.

CORE. Central part of the earth, having a radius of about 2000 miles, composed mostly of iron and nickel. The outer part of the core is molten; the inner part is in the solid state.

CRATON. The ancient, stable interior region of a continent, commonly composed of Precambrian rocks. Often covered with a veneer of younger sedimentary formations.

CRO-MAGNON. Early humans that migrated into Europe and Asia during the last ice age, about 35,000 years ago. Known for finely crafted bone and stone tools and artifacts, and splendid cave art.

DENSITY. Mass per unit volume. The density of water is 1 gram per cubic centimeter. The density of surface rocks is about 3 grams per cubic centimeter. The average density of the earth is 5.5 grams per cubic centimeter.

DIVERGENT PLATE BOUNDARY. A boundary between lithospheric plates where the plates move apart. Generally corresponds to the mid-ocean ridges where new crust is formed by the solidification of liquid rock rising from below.

EPICENTER (OF AN EARTHQUAKE). The point on the surface of the earth directly above the focus of an earthquake.

FAUNA. The animal community which characterizes a certain place and time.

FLORA. The plant community which characterizes a certain place and time.

FOCUS (OF AN EARTHQUAKE). The actual source of an earthquake beneath the surface; the place of maximum earthquake intensity.

FORMINIFERS. One-celled animals living mainly in salt water that secrete shells of calcium carbonate. Thousands of species have been recovered from oceanic sediments. They provide an especially useful fossil guide to the sediments of the past 300 million years.

GNEISS. A banded, coarse-grained metamorphic rock with alternating layers of unlike minerals; consists of essentially the same components as granite.

GONDWANALAND. Southern supercontinent of Paleozoic time, consisting of Africa, South America, India, Australia and Antarctica. Broke up into the present continents in Mesozoic times.

GRANITE. A coarse-grained, silica-rich rock consisting primarily of quartz and feldspars. Granitic rocks are the principal constituents of the continents. The source of granites has been hotly debated. They are believed to form below the surface, and probably from the molten state.

HOMINID. A member of a family of the primate order, of which only humans survive. A human ancestor.

HOT SPOT. A persistent, stationary source of volcanic activity, which remains anchored in the mantle as lithospheric plate moves over it.

HYDROCARBONS. Any of a large number of chemical compounds consisting of hydrogen and carbon. They are the principal constituents of petroleum and coal.

ICHTHYOSAURS. Extinct marine reptiles of the Mesozoic era, having a porpoiselike form and paddle fins.

IGNEOUS. Rocks that have solidified from the molten state.

IRIDIUM. A brittle, metallic element of the platinum group.

LIMESTONE. Sedimentary rock composed of calcium carbonate, generally made up in large part of invertebrate fossil skeletal remains.

LAURASIA. Northern supercontinent of Paleozoic times, consisting of North America, Europe and Asia.

LAURENTIDE (ICE CAP). The most recent glaciation of North America. The ice cap had its centers of accumulation on the Laurentide Plateau of central Quebec.

LITHOSPHERE. Rigid outer layer of the earth, typically fifty miles thick. Includes continents and ocean crust as its upper layer. Divided into segments called plates.

MAGMA. Molten rock within the earth, the parent material for igneous rocks.

MAGNETIC ANOMALY. Any local deviation from the normal magnetic field intensity.

MAGNETIC FIELD REVERSAL. A reversal of the north-south polarity of the earth's magnetic poles. Has occurred intermittantly throughout geologic time.

MAGNETITE. A dark, strongly magnetic mineral consisting of iron oxide. Sometimes called lodestone.

MANTLE. The part of the body of the earth below the crust and above the metallic core. Thought to consist of dense rocks of iron-rich silicate materials.

MERCALLI SCALE. A scale of earthquake intensity, ranging from I to a maximum intensity of XII.

MESOSPHERE. The rigid part of the earth's mantle below the asthenosphere.

MESOZOIC. Literally "middle-life." Refers to the period between 250 and 65 million years before the present.

METAMORPHIC ROCK. Rock crystallized from preexisting igneous, metamorphic or sedimentary rock under conditions of extreme heat and pressure without melting.

METEORITES. Metallic or stony bodies from space that enter the earth's atmosphere, and sometimes impact on the surface. Typically the size of a grain of sand, but infrequently can be miles across.

MID-OCEAN RIDGE. Submarine ridge along a divergent plate boundary where new ocean floor is created by the upwelling of material from the mantle.

MORAINE. A curvilinear ridge of erosional debris deposited at the melting margin of a glacier.

NAPPES. A mass of rock that has been pushed over younger formations by thrusting or folding.

NEANDERTHAL. Early humans who ranged across Europe and Asia during the most recent ice age. Sturdy and small-statured, they were well adapted for life at the edge of the ice sheets. Displaced by Cro-Magnon humans about 35,000 years ago.

OROGENY. An episode of mountain building.

PALEOMAGNETISM. The study of the earth's magnetism, including polarity and position of the poles, in the geologic past.

PALEONTOLOGY. The study of ancient forms of life, based on the fossil record of plants and animals.

PALEOZOIC. Literally "ancient-life." Refers to the period between 600 and 250 million years before the present.

PANGAEA. In Alfred Wegener's theory of continental drift, the ancient supercontinent that included all of the present continents.

PHOTOSYNTHESIS. The process by which plants create carbohydrates from carbon dioxide and water, utilizing sunlight as an energy source.

PLANKTON. Minute life forms that float and drift in the seas.

PLATE. One of the "broken eggshell" segments of the earth's rigid crust.

PLATE TECTONICS. The theory that accounts for the major features of the earth's crust in terms of the motion and interaction of lithospheric plates.

PROTO-ATLANTIC. The hypothetical body of water that separated the eastern and western continents before the convergence of those continents and the assembly of Pangaea.

PROTOZOA. Single-celled animals.

RADIOMETRIC DATING. The determination of the age of igneous rocks through a comparison of the ratio of parent elements and product elements of radioactive decay.

SANDSTONE. A sedimentary rock consisting of sand grains cemented together.

SCHIST. A finely layered metamorphic rock. Tends to split readily into thin flakes.

SEA FLOOR SPREADING. The theory that ocean floor is created by the separation of lithospheric plates along the mid-ocean ridges, with new ocean crust formed from material which rises from the mantle to fill the rift.

SEDIMENTARY ROCK. Rock formed by the consolidation and cementation of sediments deposited in low-lying regions.

SEISMOGRAPH. A device to record and measure the intensity of earth vibrations.

SEISMOLOGY. The study of earthquake vibrations. Has yielded most of our knowledge of the earth's interior.

SHALE. A general term referring to sedimentary rocks derived from muds, clays or silts. Typically fine-grained, soft and readily broken into layers.

SHIELD. Areas of the exposed Precambrian nucleus of a continent, lacking sedimentary cover. The North American shield is centered around Hudson Bay in Canada.

SHIELD VOLCANO. A broad, low volcanic cone built up by fluid lava flows of low viscosity.

SPREADING AXIS. Line of plate divergence usually associated with the mid-ocean ridges.

STRATA. Layers of sedimentary rock.

STRATOSPHERE. The upper atmosphere, beginning at a height of about seven miles above sea level.

STROMATOLITES. A finely layered structure built up by colonies of simple organisms, primarily algae. Consists of fine sediments bound into domes or pillars by the secretions of the organisms.

SUBDUCTION. The pulling down or sinking of oceanic lithospheric plates into the asthenosphere.

SUBDUCTION ZONE. An area where oceanic plate is subducted into the asthenosphere. The ocean trenches are the surface expression of a subduction zone.

TECTONIC. Refers to structures of the earth's crust formed by large-scale earth movements over geologic time.

TETHYS SEA. The hypothetical mid-latitude arm of the oceans separating the northern and southern continents of Gondwanaland and Laurasia some hundreds of millions of years ago.

TILL. Erosional debris deposited by glacial ice.

TRANSFORM FAULT. A fracture in the earth's crust along which lateral movement occurs. Common feature of the mid-ocean ridges, creating offsets in the line of spreading.

TRILOBITE. Extinct bottom-dwelling marine animals with a shell divided into three lobes. Abundant in the Early Paleozoic.

UNIFORMITARIANISM. The belief that the slow processes of uplift, erosion, subsidence and deposition presently shaping the earth's surface have acted essentially unchanged throughout geologic time.

VISCOSITY. The resistance of a liquid to flow.